TAILS, TEATS and TURNIPS

by

Jeanne M Titchiner

Grosvenor House
Publishing Limited

All rights reserved
Copyright © Jeanne M Titchiner, 2016

The right of Jeanne M Titchiner to be identified as the author of this work has been asserted in accordance with Section 78 of the Copyright, Designs and Patents Act 1988

The book cover picture is copyright to Jeanne M Titchiner

This book is published by
Grosvenor House Publishing Ltd
28-30 High Street, Guildford, Surrey, GU1 3EL.
www.grosvenorhousepublishing.co.uk

This book is sold subject to the conditions that it shall not, by way of trade or otherwise, be lent, resold, hired out or otherwise circulated without the author's or publisher's prior consent in any form of binding or cover other than that in which it is published and without a similar condition including this condition being imposed on the subsequent purchaser.

A CIP record for this book
is available from the British Library

ISBN 978-1-78623-711-8

Dedicated to
all those cheerful, hardy Women's Land Army girls
who did so much to keep Britain fed
during both World Wars and after.

Acknowledgments

Grateful thanks to
Valerie Thame
for her tireless encouragement and
creative editing
and
Andrew Barnes
for his generous technical input.

ONE

I admired myself in the mirror.

Pork pie hat – khaki, perched saucily over one eye.
    Dark green sweater over a beige aertex shirt – with a tie I didn't yet know how to put on.
    Smart khaki-coloured breeches, nipped in to my small waist with a brown leather belt.
    Knee-length socks to match, with brown lace-up shoes – a bit *too* sturdy.

In uniform at last! A newly signed-up member of the Women's Land Army.
    On my bed another two parcels held less glamorous garments – a pair of beige dungarees with short matching jacket, another aertex shirt, a second pair of socks and some gumboots. A note enclosed told me that a winter overcoat and a waterproof were not available at present, suggesting I supply my own in the meantime. I hoped to get them soon; I'd had to hand in most of my precious clothing coupons – not enough left for a new dress for Christmas.
    World War II had finally ended in August and the W.L.A. Representative who signed me up said materials were in very short supply, as was everything else. I'd answered the urgent call of our new Prime Minister, Mr Attlee, for more women volunteers to work on our farms. Most men weren't back from the war and our country was more desperately short of food than ever.
    In my minds-eye I saw him, sternly pointing a finger at me – like Lord Kitchener in the World War I poster saying **'Your Country Needs YOU'**. How could I resist?

For a healthy, happy job

Anyhow, I'd been impressed by stories a friend told me about her happy Land Army life; driving a tractor, living in a hostel full of friendly jolly girls, going to nearby camps for dances and entertainment – delights my naïve seventeen-year-old self could only imagine and long to experience.

But first there were obstacles to overcome. My Mother shrieked in dismay, begging me not to join, urging me to 'get a nice shorthand and typing job in the engineering firm down the road'. She'd lost one daughter to a handsome American G.I. and the thought of losing the other was too much.

After all, I'd just finished a shorthand and typing course which I'd quite enjoyed – cherishing dreams of one day becoming the Prime Minister's Confidential Secretary and sharing all the State secrets.

But before that my life-long ambition had been to go to Art College and become a famous fashion designer, though this had been cruelly thwarted; my now permanently absent father wasn't inclined to support Mother and me anymore. So this third career craze was too great a shock for her.

Restless, determined to have my own way, to leave home and taste adventure, I secretly found the nearest W.L.A. Recruiting office, who sent me for a medical. I passed with flying colours. 'A fine healthy girl; fresh air will be the making of you,' the doctor said. I wondered if he said it to every hopeful, so desperate were they for recruits.

And so it was that on the 5th November 1945, proud but self-conscious in my smart uniform, I was collected by a matronly, tweed-clad Mrs Bartholomew and taken by car to Brinkleigh Farm, near Fleet in Hampshire. When I'd first heard I was to work on a small dairy farm, just me and three men – no hostel with other girls – I was disappointed and rather scared, but when we arrived at the farm gates my spirits revived.

Down a short drive was a pretty, thatched and beamed Tudor cottage. In the fields behind I noticed a herd of grazing cows and a little red tractor chugging along.

Immediately, I thought of the posters I'd seen of happy, glamorous land-girls holding pitch forks and milking pails, with kindly, rosy-faced farmers benignly looking on, against backgrounds of sunny fields, cows and corn sheaves.

Everything's going to be wonderful, I told myself; this dream is coming true.

'Hurry along, dear,' Mrs Bartholomew said, 'Mustn't keep Colonel Fortescue waiting.'

Meeting the Colonel was an ordeal but I needn't have worried. He reminded me of my father I missed so much, with his military bearing; handsome, moustachioed, with a comforting aroma of cigars and a welcoming smile.

Introductions over, he ushered us into the sitting room where his wife was reading the newspaper.

'Here's the young lady, Monica, who's come to help Larchford with the milking.'

Barely lowering the paper, she looked at me dispassionately, gave me a half smile and a slight nod.

'How do you do, Miss Parsons. You'll be sleeping here, but having your meals with our cowman and his wife. My husband will show you your bedroom.' She disappeared behind the paper.

Mrs Bartholomew fussed her goodbyes, promising to call in a few weeks to see how I was getting on. I suddenly felt alone, missing Mother. Mrs Fortescue obviously wasn't going to be a second mum to me.

Colonel Fortescue took my case and I followed him upstairs.

'This is yours,' he said, throwing open a door to the prettiest bedroom I'd ever seen: rose carpeted and curtained, with a dainty dressing table and large comfortable looking bed.

'Oh, it's lovely,' I said. A bedroom of my very own for the first time in my life – a thousand times better than sharing a bed with a Mother who snored, up in granny's attic where we'd finally taken refuge in the war. As an Army child, first in India, then London, and as a reluctant, often unhappy evacuee, I'd slept in a motley assortment of beds, but this looked truly luxurious.

'Glad you're pleased, girlie. The bathroom next door is all yours too,' the Colonel said, patting me on the shoulder. 'Come down when you've unpacked and I'll take you to meet my herdsman, Larchford. He'll show you the farm.'

More luxury! My own bathroom, with a heated towel rail and I could have a *daily* bath; At home it was once a week only on Fridays and a lavatory at the bottom of the garden!

The large cottage stood a field away from the farm buildings, via a well trodden path. Again, I suddenly felt alone and shy to be meeting the man who would be my real, day-to-day boss.

We found him in the dairy yard and as he turned at the sound of our voices, I got the shock of my life; a tall, thin man with an aquiline nose, a pale complexion and a pair of startlingly blue, piercing eyes which he turned on me, unsmiling, seeming to size me up.

'I'll leave you with Larchford, young lady. He'll show you round and take you over to the Baggs' place. Okay, Larchford?'

Larchford nodded, still staring at me. Uncomfortable, I didn't know what to say.

'Huh! You're a bit young,' he said. 'Any dairy farming experience?'

I shook my head.

'Done any farming?'

Again I shook my head.

'Gawd help me! What are you doing here then, Fanny?'

'My name's not Fanny. It's *Jeanne* Parsons. I, I like being outside – and animals.' (I didn't say I'd only ever known dogs and cats)

'What a bit of luck! You'll be seeing a lot of those, especially rear ends. Call me Jim, do as you're told, learn quickly, keep your nose clean and maybe you'll do, Fanny.'

'*Jeanne*,' I repeated, indignant.

'*Fanny!* Follow me, quick sharp!'

My spirits sank to the soles of my highly polished, sensible shoes. Where was the kindly, benign farmer of my imagination? Jim Larchford wasn't even a country man. He came from London. I could tell from his accent because I'd lived there before the war.

'What the bloody hell *can* you do?' he asked, as I hurried behind.

'Shorthand and typing – oh, and I like painting and singing.'

He groaned. 'That'll come in very handy, Fanny.'

'*Jeanne!*' I said, resenting his sarcasm and beginning to dislike him.

'The milking shed!' he said, throwing open a stable door. 'Gascoigne six-pen electric milking machine – just installed.'

I gasped at the sight of a bewildering array of meandering glass tubes, large bottles suspended on what looked like weighing scales and other extraordinary unidentifiable objects – and *I* was going to have to learn to operate them! I'd imagined myself hand-milking, head pressed against a cow's soft flank, listening to the rhythmic sound of warm milk into a bucket, perhaps even humming a little tune.

'You can watch me tomorrow morning, learn how it's done.'

Speechless, panicking, I followed him round the rest of the farm buildings in a daze, trying to keep up and take in what he told me.

'Milking and the cows will be your main job, no pig work; helping my wife sometimes with calves and chickens; then it's all hands at haymaking and harvest time. We'll soon lick you into shape – I hope to God!'

He strode off towards the field of cows and as we got nearer they looked up and began walking towards him. I noticed they were smallish with pretty faces and big limpid eyes. I'd never been so close to cows before. Cautiously, I stepped behind Jim.

'About forty milking Jerseys, pedigree attested, and one Shorthorn, Alice, also attested.'

Alice looked enormous as she barged her way to the front and nuzzled Jim's hand, looking comically incongruous with her dainty cousins, who were all vying for Jim's attention. I hadn't realised cows smelt so strongly.

'Why do you keep Alice?' I asked.

'You may well ask; she's a left-over from the last herd, aren't you lass.'

In a field beyond I spotted five or six horses, two very large.

'Do those belong to Colonel Fortescue?' I asked.

'Yep, two working carthorses, Molly and Dolly; the little pony pulls the trap, the others are ex-polo ponies – the Colonel used to play, but nobody rides them now – pity. Come on, Fanny, follow me. Let's get you over to Alf Baggs; his missis might give us a cuppa.'

The Baggs lived in a modern, semi-detached bungalow in front of the farm buildings. As we arrived, Alf Baggs came out of the back door.

'This is the girl we've been landed with, Alf,' Jim said. 'She don't know nothing – no training. It's up to us.'

Alf stared at me disapprovingly from tiny round eyes set in a shiny pudding face and opened and shut his button-sized mouth by way of greeting. He reminded me of a currant bun.

'How do you do,' I said stiffly, (these country people didn't have any manners!).

With a nod he carried on towards the farm. Jim knocked on the door, to be answered by an exact replica of her husband.

'Doris, this is Fan – whoops! *Jeanne* Parsons – you're going to feed. Hope she don't eat you out of house and home!'

'So do I,' she answered, looking at me hard. 'What with rationin' gettin' worse. Don't think, miss, I'll be cookin' special for you; you'll eat what we do and like it!'

'Thank you,' I muttered, my spirits sinking to Australia. Why was everyone so horrid when they should be grateful I'd come to help out?

'Too late for tea, miss; supper's at six, sharp mind you. I don't hang around for no one.'

'Don't go getting on the wrong side of her, Fanny,' warned Jim, as we walked back to the farmhouse. 'She's got a tongue on her, all right!'

Thoroughly miserable, I went up to my room, deciding to have a really hot bath to cheer myself up before I braved Doris Baggs again.

Supper was really a tea. Bread and margarine, jam, fish paste and a thin slice of rather dry cake; not very satisfying and eaten in uncomfortable silence, except for Peggy, their little roly-poly daughter, who seemed eager to chat to me about school.

Ordeal over, I fumbled my way back to the cottage in darkness, vowing to buy a torch when I next went home.

Disheartened, near to tears, there was nothing to do but go to bed. I had to be in the milking shed by five forty-five, so set my alarm for five a.m. I looked out my dungarees, boots and jacket, then lay awake, my mind full of conflicting feelings and apprehension, longing for oblivion.

* * * *

TWO

Bleary-eyed from lack of sleep, I staggered across to the milking shed in the gloomy dawn. The yard was already full of impatient cows, milling around the milking shed door, anxious to be first in. The lights were on inside.

Where was the cup of tea Mother always brought me before I got up? My tummy rumbled. When would I get some breakfast?

'Blimey, Fanny, been out on the tiles all night?' Jim cried. 'Chop chop, let's get started.'

'But the yard's full of cows,' I wailed. I'd have to walk through them.

'*So?* I'd be really worried if it was full of sheep. Just push your way through, you spineless nitwit!'

It was the bravest thing I'd ever done, but my fear of Jim was greater than of these restless, head-tossing creatures. As I timidly eased myself amongst them to the door, they all surged forward, nearly knocking me down, but Jim shooed them back and I fell in, shaking and breathless.

'A good start, Fanny. Now, stand in that corner by the record sheets and listen to what I tell you –just *once*, mind you!'

I was only too pleased to squeeze into the farthest corner. I wished I could vanish!

'Right, I've already got the cows up from Long Meadow, turned on lights, switched on machine, checked pulsaters are ticking, bottles and teat-cups okay. Three pails of hot water and clean udder cloths at the ready – got that, Fanny?'

I nodded furiously, certain I hadn't 'got' anything.

'Through to the dairy, assemble the cooler – never forgetting to roll a churn under – okay?'

'I think so,' I whispered, though my mind was beginning to spin.

'All ready now,' Jim said. 'We'll let the girls in.'

I flattened myself against the wall as he opened the door to let in the first eager six. They might just as well have been lions, I'd never felt more terrified. With noisy mooo's his girls quickly found a stall each.

Jim came over and looked through sheets of paper on the shelf near me.

'Ministry of Agriculture's daily reports. According to each cow's yield the day before, they get an allocation of mixed feed. You enter the yield by each cow's name.'

'*Name*! How can you tell which is which?' I asked, astonished. 'They all look alike. I'll never tell them apart.'

'Have I got a blitherin' idiot here, I ask myself? You'll learn!' He pumped a handle by each stall and from a hob by the cow's head, feed came tumbling into a trough.

'Now, I'm washing each cow's udder first and dipping the teat-cups into the pail; all clean. Come over here, Fanny. Watch how I put the teat-cups on diagonally, front left, back right, front right, back left – got it?'

I nodded again, desperately hoping I'd remember things in the right order.

The teat-cups hissed as they were sucked on and began to pulsate rhythmically. Unconcerned, the cows enjoyed their breakfast. Not like me, totally concerned and longing for mine.

Immediately, creamy milk frothed up in the bottles, the scales beginning to register the output. Magic – but oh, how complicated! Wouldn't it be easier hand milking, I wanted to ask?

'How, er, how d'you know when they're finished?' I asked.

'Feel their bleedin' udders, dozy girl and remember yesterday's yield.'

It didn't seem possible to watch six bottles and udders, dart backwards and forwards to enter the Min.Ag. sheets, remember each cow's name and yield and let them in and out.

I almost smiled when Alice barged in, gave me a very curious look, then squeezed herself into an empty stall. Her teats looked enormous compared with her Jersey cousins; her milk whiter, her yield bigger.

As Jim let each cow out, he turned a lever, sending milk rushing through the glass tubes above to disappear into the dairy beyond. Without a pause he was letting more cows in, one and two at a time, washing udders, checking yields, turning handles. It was fast, non-stop work with not a second to think what to do next. I felt sick – you even had to remember to change the churn a couple of times. My brain was almost in melt-down.

I sighed with relief when Jim let the last cow out, thinking we'd finished, but oh no, another shock – cleaning up!

Just as Jim was showing me how he did it, a back door opened and a tiny, dark-haired woman wearing huge glasses that almost hid her face came in carrying two mugs and a large jug.

'Tea – bet you're ready for it, gal,' she said, smiling at me – the first smile I'd had since Colonel Fortescue's welcome. I immediately liked her, so in need was I of a friendly, smiling face.

'Guess you're the new girl,' she said, pouring out the tea.

'It's Fanny Parsons, Masie. Come to take over my job so I can sit indoors toasting my toes while she does the milking.'

'I'm *Jeanne*.' I told her, glaring at him.

'Oh, take no notice of him, he just likes stirring,' Masie said, as I gulped down the best cup of tea I'd ever tasted.

'You're holding us up, Masie,' said Jim. 'Got work to do. See you at eight.'

'Don't you be too hard on her, Jim, or you'll answer to me,' Masie warned him, giving me such an understanding smile I wanted to cry on her shoulder.

Jim grunted, and showed me how he washed through the machine and in the dairy, dismantled the steel cooler, then plunged the pieces into a deep metal sink full of extremely hot caustic-soda water. He handed me a big scrubbing brush.

'Clean that lot, Fanny, scrub hard; I'll see to the churns – show you how to roll even the heaviest tomorrow – a doddle! After the cooler, reassemble it, wash the cloths and scrub out the buckets.'

The hot caustic-soda water stung my eyes and scalded my hands as I clumsily wielded the big scrubbing brush.

'Put your back into it, girl, we've got the floor to do yet.'

Grabbing two other pails, Jim filled them with cold water, and after impatiently helping me re-assemble the cooler, we washed down the milking shed. He showed me how, with a flick of the wrist you could cover a wide area of floor with water. We scrubbed as we went with huge brooms, sweeping away into a gully the hoof mud and large cow pats – Jim's girls were not house trained!

By now my back ached, my head ached and my stomach groaned from lack of food. It was almost eight o'clock and I'd been up since five!

'That'll do, Fanny, get off for your nosh. Don't keep madam waiting! Back here before nine, mind you, and we'll take the cows up to Long Meadow.'

Breakfast with the Baggs was as silent as last night's supper with just enough porridge, bread-and-scrape and jam to suit an office worker but not a starving dairy maid! Alf kept examining me over the top of his paper as if I was some strange breed of animal he didn't like. I thanked my stars for Peggy, cheerily chatty as before.

After my meagre breakfast, I rushed back to my room, on the way up noticing a framed photo on the stair wall of a beautiful young woman gorgeously dressed for Presentation at Court to the King and Queen. Looking closely, I realised with amazement it was Mrs. Fortescue! I couldn't equate this lovely debutante with her present self – dull, untidy hair, pale unmade-up face and scruffy clothes – a faded beauty. I'd even seen her at bedtime trailing along the landing in a tattered, crumpled dressing gown. She didn't know about mending for sure! Shocked, I vowed I'd never allow myself to get like that, intending always to wear lipstick and powder when I did the milking and do my mending as I'd been taught.

It was quite pleasant on this dry November day, walking the cows up the lane with Jim to a really long field beyond a big bungalow, which he said was his.

'Time you met the fellas,' Jim said. 'Follow me.'

Thinking he meant other farm workers, I was startled to find myself staring into the faces of two malevolent-eyed bulls, glaring at me over their stable doors.

'Meet Romulus and Remus – he's the old man, semi-retired; Romulus does the business mostly.' Jim told me.

I didn't know what 'the business' was, but prayed it didn't involve me, especially when Romulus let out a huge snort and stamped the floor, while Remus crashed his enormous head again and again against the door, threatening to break it down!

'Right! said Jim, 'They need exercise while we clean out their pens.' Grabbing a long pole with a clip on the end, he attached it to Romulus's nose ring and calmly opening the stable door, led him outside. 'You must hold his head fairly high, see– then you've got control. If you lower it, he could charge.'

'Me!' I bleated. 'You, you want *me* to hold him?'

'You either do that, you wimp, or dung out the bloody pen – WHICH YOU DON'T KNOW HOW TO DO!'

'Don't shout,' I dared to protest. His language shocked me!

Grunting, he thrust the pole into my trembling hand, pushing my arm up to hold Romulus' head higher.

'Lead him forward a bit,' Jim ordered, fetching a wheelbarrow and spade from a nearby shed.

I trembled all over as Romulus snorted angrily and pawed the ground.

How was it that every time I thought I'd reached my pinnacle of fear, another terror presented itself?

'Hold him still, girl and see how I dung out.'

With one wary eye on Romulus and the other on Jim, I watched him scrape up the acrid-smelling straw, spading it into the wheelbarrow at incredible speed, then tipping it onto a nearby heap and repeating the process again and again.

'Got that, Fanny?' he said, bringing out a bale of straw from the shed and using enough to re-cover the concrete floor; finishing off by replacing Romulus' drinking water. He took the bull-pole from my quaking hands and walked Romulus round the field once, while I struggled to recover my nerve.

'That's better boy, isn't it,' Jim said, patting him affectionately and leading him into his clean stall. 'Now it's yer uncle's turn.'

I almost fainted. 'Uncle' was twice as big and ten times nastier-looking than Romulus.

'I'll exercise the old man,' Jim said. 'You get dunging out!'

I nearly vomited as I wielded spades-full of wet, revolting straw into the wheelbarrow, then dismayed to find it too heavy for me to push, however hard I tried. Jim was nearly back from walk-abouts with Remus. I waited for the inevitable.

'Good grief, Fanny! Call yerself a land-girl? You'll have to toughen up damn fast or you're no good to me. Here, hold the old man – I'll finish the job.'

I sensed Remus' utter contempt for me as I clung to his pole with sweaty hands, while he bellowed and stamped.

My self esteem was at zero when Jim and I met Masie, carrying two buckets.

'Hello, long faces!' she greeted us. 'Just going to feed the chickens and collect the eggs. Let her come with me, Jim, it's nearly dinner time.'

Being with Masie was a tonic, with her cheerful chat while she scattered grain amongst her clucking charges and we collected eggs from their coops. If only this was my job!

I badly needed the hour off Jim gave me. After devouring a dinner of greasy stew and potatoes, followed by lumpy semolina – eaten in silence with no Peggy to talk to, I rushed up to my room and threw myself on the bed.

I daren't even doze – Jim had said I'd be helping with the afternoon milking as it was Alf's day off. I knew I'd be expected to remember everything and in the right order. I grabbed a pencil and paper and scribbled down what I recalled Jim doing and how he did it, but my brain was not co-operating.

I was only half way through my first day on the farm and it already felt like a nightmare year!

****

THREE

At two o'clock sharp, Jim took me across fields to where the rather shabby red tractor was busy; harrowing, he told me.

'Keep up, Fanny,' he said. 'Want a word with Gummer. I need all hands to build the cows' wintering-in yard before it gets too cold. Can't miss this dry spell.'

'Can I drive the tractor?' I asked hopefully. It looked far safer than handling Jim's fellas.

'No you can't!' he snapped. 'You haven't even learned to milk the ruddy cows yet.'

The tractor put-put-putted to a halt and a friendly, freckled-faced man with an abundance of red hair, jumped down.

'Ah, you're the new milk girl; I'm Bert,' he told me with a grin. 'How you doin'?'

Before I could open my mouth to reply, Jim broke in 'Don't ask – she *isn't!* How are you fixed tomorrow for getting the rick down and building the winter yard – got to catch the weather.'

'Umm, providin' I can finish this field by dinner time t'morrow, I'll give you a hand, right?'

'That'll do,' Jim nodded. 'I'll get Paddy to help out. Come on, Fanny.'

'Who's Paddy?' I asked.

'Looks after the Colonel's horses; cushy job if you ask me,' Jim said. 'Lives over the stables.'

Without knocking, Jim threw open the door of a rather ramshackle wooden building and yelled up the stairs for Paddy.

The place smelled of polish and old leather and hanging from pegs on the walls were several rather dusty saddles with festoons of harness and other bits and pieces.

'Sure, I'm comin', I'm comin',' a voice called from above, and a plump little man with a Father Christmas beard and bushy eyebrows appeared, looking as if he'd just woken up.

Not bothering to introduce us, Jim asked him brusquely if he'd help out the next day.

'Well now, let me see,' Paddy considered, not looking too keen I thought, 'if I've the time after feedin' m' horses and other little jobs, I could spare an hour, I'm thinkin'.'

'I'm thinking you'll spare more than an hour, man – two o'clock sharp,' snapped Jim, turning to go.

Paddy made a face at Jim's back, and winking mischievously at me, mouthed 'Come an' see me, missie – any time.'

'Bloody Irish layabout,' Jim muttered, 'spends nearly all day on his bed, tipping back the whisky.'

I rather liked Paddy and was determined I'd visit – when I got some time off.

My stomach flipped right over when Jim said, 'got to get the cows for milking.' My moment of reckoning was fast approaching!

All the way up the lane, I silently muttered to myself, trying to remember the sequence of everything I'd have to do. I willed myself to stay calm, but it felt worse than taking exams.

'Well, Fanny, the cows are here. What next?' Jim said, folding his arms.

Easy!

'Switch everything on,' I said – but couldn't find a single switch. Jim pointed behind the door. What a daft place!

'And?' he asked.

'Check the machine and the dairy.' I felt pleased with myself when he nodded. Glancing along the ticking, humming machine, it looked okay – though I wouldn't have known if it wasn't. I dashed through to the dairy and checked the cooler was set up and Jim showed me how to roll an empty churn underneath.

'Ready for the cows?' he asked.

'Yes' was on the tip of my tongue, but in the nick of time I shook my head and rushed around filling the sink and preparing three pails and cloths. Jim told me how much caustic-soda to throw into the sink.

'Not bad, Fanny. A bit slow, but I'll make a land-girl of you yet.'

Praise at last. My confidence rose. So far so good. I was going to be fine.

'I'll let the girls in,' Jim said, but at the sight of those impatient, single-minded cows, my courage began to wane. As if their front ends weren't intimidating enough, I had yet to learn how to handle their back 'business' ends, as Jim called them.

As each cow pushed in, he called out their names; 'Princess, she's always first; Rosa's got white legs; Daisy's got a flower-pattern between her ears; Bella's nearly all brown – on and on, as if I'd be able to remember even *one* – except Alice!

Tentatively washing Daisy's udder and teats, I felt I ought to ask her permission first; it seemed such an intimate intrusion!

But Jim shouted, 'Get stuck in Fanny, they won't break off.'

I was all thumbs and indecision that first time fixing on teat-cups – was it right front, left back first – or the other way round? Whatever, they fell off and I had to re-wash them and try again. Success this time!

I hadn't bargained on fidgety hooves and flicking, dung-splattered tails, until Jim told me to always 'feed the buggers first' to prevent a mutiny.

'Get a move on, slowcoach; we don't want to be here 'til midnight,' he shouted.

The more I dashed from bucket to udder to record sheet to the next cow, fumbling with handles and struggling with levers – not often in the right order – the more Jim barked orders, his colourful expletives raining down on me like hailstones, and the more muddled I got!

'Stop flappin' about like a bloody windmill, Fanny,' he yelled, coming to my rescue. You're scaring my girls.'

When Alice came in, late, she frightened me rigid, pinning me up against the wall and sniffing me all over, not sure she approved of this new entity in her domain.

By the time the last cow left – Duchess, an old lady whose udder almost brushed the ground and Jim said was always last, I was mentally fragmented and nearly asleep on my feet as I scrubbed the utensils and we cleaned the milking shed floor.

'Five thirty sharp tomorrow morning, girl. We've got to build the winter yard between milkings.'

I didn't even have the strength to nod and staggered back to my room to wash, with only ten minutes to spare before another meagre, silent tea, Peggy having been sent to her bedroom early.

Collapsing into my bed that night, every tiny muscle and bone in my body throbbed and ached.

What had I let myself in for? How would I survive? What could I tell Mother when I wrote?

Mother! How I longed to be at home with her – and to think I'd wanted to leave so badly.

Eventually I fell asleep ..... to find myself crouched in an enormous bucket spinning round and round in a fast-running river of frothing, creamy milk. Bright green teat-cups dangled from overhanging, sharp-thorned trees. Through the glass branches peered the huge red faces of evil-looking bulls and wild-eyed cows, bellowing and mooing. The bucket spun faster and faster. I could see rapids and a huge waterfall ahead. I tried to scream. I was about to go over – but a shrill ringing sound woke me!

It was four forty-five a.m. Time to get up.

****

FOUR

The trauma of the next few days was relieved only by sudden joyous revelations – a cow named; an emptied udder; a churn changed, yesterday's yield and today's feed ration remembered. Not that Jim acknowledged these eureka moments. He'd decided I was 'the most spineless nincompoop he'd ever had the misfortune to be lumbered with' and I was 'still tied to mummy's apron strings'!

I guess he had a point!

Silent as ever, Alf helped with afternoon milking, it should have made things easier but his resentment of me was blatantly obvious – why, I couldn't imagine.

The only boost to my deflated ego was from Masie when she brought over the oh-so-welcome morning mug of tea, or if I bumped into Colonel Fortescue, who'd ask about my progress, patting me encouragingly on the back, though I never told him Jim's opinion of me!

My difficult letter to Mother I kept non-committal; praising my lovely bedroom, Colonel Fortescue's kindness and just saying I was learning to use a milking machine. I said I didn't know when I'd get time off, but I'd come home when I could – *'come home and stay home'* was what I longed to say!

The wintering-in yard took two days to complete; dismantling a straw rick and heaving the bales into position to create medium-high, three-sided walls butting up to the back of a barn near the dairy.

My efforts to pick up a bale and heave it onto my back ended in deep humiliation, with Jim and the others laughing hilariously as I lay sprawled on the ground under my load.

'You'll get the hang of it, Fanny,' Jim said, and thrusting a stick in my hand, told me to fetch the cows up from Long meadow ready for milking – a job I was dreading – but then I met Colonel Fortescue by the farm gate and he insisted on coming with me, for moral support.

He told me to call at his office every Friday afternoon for my pay; the vast sum of sixteen shillings a week, plus my keep. For three shillings a time, Bert Gummer's wife would wash my dirty work clothes, but not underwear.

The cows safely in the yard, I appreciated the Colonel's usual reassuring pat on the back.

With the wintering-in yard ready for cold weather, Jim took me, and the farm cat, up to a warm, sweet-smelling loft built over the milking shed, full of sacks of cow feed – oil cake and cereal concentrate, he told me. From these sacks, the large hoppers over each cow pen below had to be filled every two days. This was to be my job.

I soon found out why he'd brought the cat, who was scrabbling to be put down.

'Nice treat for you, Tiger,' Jim told him, pulling out the sacks.

I gasped as several mice ran out, leaving little nests of minute, hairless babies – so sweet, so helpless.

Immediately Tiger leapt on them, crunching each one down in great mouthfuls. I was horrified!

'Stop him, stop him!' I wailed, bursting into tears.

But Jim was laughing. 'Oh dry up, Fanny, they're pests; chewing through m'sacks, eating precious feed. Go on Tiger, fill yer belly.'

Disgusted and angry I fled downstairs. That was not going to be my job. I'd refuse, however horrid Jim was to me!

Feeling mutinous, I marched off to see if Paddy was in.

'Welcome, welcome, missie,' he said, 'I was hopin' you'd come by soon. Sit you down. Would you care to join me in a glass of somethin'?'

'Water, please.' I said, wondering what a 'glass of somethin'' was? Better play safe.

I suddenly didn't know what to talk about, but needn't have worried. Paddy launched enthusiastically into his life history, how his family came from Killarney. Way back, he'd married and had a son, but lost contact.

'I was a jockey in m'youth, a good one, sure I was. After that, groom to the Colonel's polo ponies. More's the pity he don''t ride now. But him hangin' on to them gives me a job, so I'm not complainin', I'm not.'

'Why did you come to England?' I asked.

'You go where the work is, so you do!' he said. 'Now, missie, I'm taking the trap into Fleet this afternoon; it's fine days we're gettin' for November. You come with me, get away from that bully Larchford.'

'Oh, Paddy, I'd love to, but it's milking soon. You could post a letter for me though.'

'For sure I will, and you'll come another day?'

I nodded and rushed off to get my letter to Mother. I could think of nothing nicer than bowling down a country lane in Paddy's trap and wished it could be today.

Unfortunately, I bumped into Jim on my way back from Paddy's.

'So that's where you've been.' he said. 'What bullshit has he been telling you?'

'He's posting a letter for me,' I said in an offhand voice.

'Don't you go all hoity-toity with me, girl! Come on, we've got milking to get on with.'

Whether it was anger I still felt over the mice, I don't know, but I marched into the dairy with my chin in the air and a determination to just do my best and answer back if he got too horrid!

It worked! Somehow I remembered more, recognised more cows, gained a little confidence and stood up for myself a few times.

Another miracle was happening. I began to realise the cows not only *didn't* look alike, but had personalities – just like humans!

Princess had a spiteful streak, head butting her way to the front and aiming a surreptitious kick at me if she got the chance, or stepping on my toes. Daisy was docile – a bit thick even; Alice always gave me a superior look as she pushed into her stall and dear old Duchess was just that – a peaceful, good-natured old lady content to take life slowly.

I was astonished to find, when I talked to them, they responded and knew their names, and one cow, intelligent, expressive-eyed Josephine, loved to be petted. She was to become my favourite and we were to have lots of close conversations in the months to come.

****

with my gentle Jersey girls

FIVE

November continued surprisingly mild, with Autumn leaves still floating lazily down, and I began to fall into the rhythm of moving cows around, milking, scrubbing, lugging pails of water, dunging out, (only ever able to push a half-full wheelbarrow), timorously leading Romulus and Remus around, re-filling hoppers – without Tiger – and occasionally helping Masie with the chickens; my favourite time. Then straight after tea, still hungry, collapsing into bed aching all over.

Jim shouted at me less – I'd got the hang of milking – but he never praised and now he had a new goal. Speeding me up. He'd count out loud how many minutes I took to do each task and it was always too long. He was especially infuriating when I had to go up into the loft to re-fill the hoppers.

'Get a bloomin' move on, Fanny,' he'd yell. 'One minute *only* to get up those stairs – you're only half way up. One minute to open up the sacks and hoppers. You should be filling'm now – *two* minutes, but good grief, girl, you haven't even started, have you!'

I fumed. Was there no pleasing him? He was like one of those awful 'time-and-motion' foremen in factories – not a normal human being!

I certainly felt I'd earned my wages on Fridays and at last Jim said I could have the following Sunday off. The outside world had faded, but now I was going home!

Examining my face in the mirror, I was shocked to see Monica Fortescue gazing back at me. Dirty, straggly hair, rough-skinned face, no powdered nose, no lipstick and my hands red-raw with broken nails. I couldn't go home like this.

I washed my hair, creamed my face and hands, then took my dirty work clothes round to Bert Gummer's wife, thinking they'd be clean by the time I returned on Monday.

Ethel Gummer proved to be as tall, angular and talkative as her neighbour, Doris Baggs was short, round and silent.

Snatching my proffered three shillings and the clothes, Ethel's darting eyes made an immediate inventory of me, while she riddled me with very personal questions about my parents – where had I come from and how did I like working with Mr. Larchford – a question I managed to skirt around. I'd have to watch what I said to Ethel Gummer.

The tediously long and scary walk to Fleet Station after dark on Saturday evening persuaded me I'd bring my clanky old bicycle back. The rail journey home to Basingstoke took far less time and as I stepped off the train in my smart uniform, my old happy self bubbled up. I looked forward to Mother's pampering and seeing my friends at church next day.

Living in an all-female 'Victorian' household with Mother and her three sisters, where the only men ever to cross the threshold was the Vicar or a curate for Sunday tea, meant it was an extremely quiet, ordered regime. With Jim's expletives still ringing in my ears, I felt quite disorientated. I daren't mention things like udders, teat-cups or dunging out and chose my words carefully in answer to the many questions I got asked.

Mother fussed over me wonderfully. For the first time ever, I really appreciated her, and my Aunts' home cooking, and pleased them by telling them so.

My smartened up appearance in church next day had the desired effect and little did my friends realise what I now knew about my life in the Women's Land Army!

Returning to Brinkleigh farm very early Monday morning was like re-baptism by fire, with Jim shouting and timing my every move, Alf Baggs silent disapproval, Doris's measly meals and hardly the energy to bath before crawling into bed.

Bravely resolving to do something about my hunger, I uttered unforgivable words. Like Oliver Twist *I asked for more*, courageously telling Doris Baggs I thought I'd be getting extra rations as a land-girl.

As I rather expected, she refused to feed me any more, saying I was greedy and a bad influence on Peggy – a total lie!

Mrs Fortescue sent for me and told me in future my meals would be provided by Masie Larchford.

I was horrified. The thought of enduring Jim every hour of every day without a minute's peace was unthinkable, but what could I say? I had to eat somewhere.

So it was with great trepidation I presented myself at the Larchford bungalow for tea. All Jim had muttered was 'So you're eating with us. Six to half-past, Masie says.'

My first surprise was to realise they had two children, Billy, eleven and Susan, nine. Instantly I felt pity for them. Fancy having a Dad like Jim; I bet he gave them hell!

Masie greeted me warmly and sat me by the fire opposite Jim, looking so out of place sprawled in a chair, wearing a pair of old slippers. The children's friendly chatter was the only thing covering up my awkwardness.

Maise's tea was *supper*; soup, chunky bread with cheese and hot jam tart. Everyone tucked in and Jim – a transformed character! Friendly, funny and affectionate to his kids. I was dumbfounded! After clearing up, we all played a board game and nine o'clock came before I realised I was tired. Jim walked back as far as the farm with me – he always checked the animals – and I had my torch to light me across to the farmhouse.

If I hoped I would now have a new, friendly relationship with Jim, I was so wrong!

There were two Jims. My tough, exacting boss and Billy's and Susan's jolly Dad. The two rarely mixed!

****

SIX

Winter had really arrived by late November, with temperatures dropping and a light dawn frost making me much more reluctant to pull myself out of a warm bed and stagger across to the Milking shed.

Just in time, I received my thick overcoat, but no waterproof – I'd have to make do with my old school mack I hoped I'd never have to wear again. Mother knitted me some very thick gloves and a long scarf, but for wet jobs and milking cows they were hopeless, so couldn't save me from frozen fingers and ears.

Each day seemed split into two. The warm, happy meal times with the Larchfords and the hard, anxious hours I spent under Jim's all-seeing, critical eye.

My expertise in the milking shed had platformed. The cows at last became recognisable personalities with names I now remembered, but Jim stopped me singing to them – 'they're here for milking, not a bloomin' sing-song!' According to him I was still either 'too bloody slow' or 'leaping about like a ruddy ballet dancer' trying to catch up.

Watching silent Alf, he got away with doing everything at the same slow, plodding pace, Jim never urging him on even once. I felt quite resentful.

By now the 'girls' were cosy in their winter quarters. Not far to get them for milking, but lots more dunging out to do – my back ached just thinking about it! They also needed regular hay feeds now they couldn't graze, and their water troughs topped up daily.

I'd become quite expert at rolling full milk churns to the back dairy door ready for lorry collection, not forgetting first to transfer some of the rich, creamy milk to a smaller churn for the farm's consumption.

I went home on another day's leave, though being a weekday couldn't go out with my friends, but still enjoyed Mother's cosseting and the quiet house. I was issued railway warrants for these home trips – an unexpected help.

Imagine my panic when I got back to Brinkleigh to hear Jim say that from tomorrow Alf and I would be doing the afternoon milking on our own. I dreaded it and guessed Alf wasn't pleased, though Jim obviously thought I was experienced enough now to be trusted.

From the start, Alf was purposely unhelpful, so I tackled him two days later.

'Alf, why don't you like me? You're acting really difficult.'

He looked taken aback and just grunted, but I repeated my question, determined he'd answer.

'Uh, well now,' he muttered, 'I don't hold on you young flibberti-gibbet girls messin' about on farms, not knowin' nothing, doin' men's work. It's not right. There, I've said it.'

So now I knew – and that was the most he ever did say to me.

Although his stand-off was to continue for the rest of my working life at Brinkleigh, he was slightly less un-cooperate in the months to come.

Mrs Bartholomew paid an unexpected visit one day, asked me lots of questions which I answered tactfully, spoke privately to the Fortescue's and seeming satisfied went on her way. I'd surprised myself when she asked me if I was happy here, to reply 'Yes', suddenly realising I didn't yearn to be at home any more! I wanted to get better at being a land-girl. To go through a whole day without Jim shouting at me. That would be success.

I whispered all this to Josephine when I tucked her up at night with her sisters in their cosy winter shed.

When I got an odd half-hour off between jobs, I'd visit Paddy and confide in him. Warmly welcoming, I got offered ' a little glass of somethin', but always settled on a mug of strong tea. I didn't care that he told me highly unlikely stories of early exploits and glorious racing victories; they were exciting and he took a genuine interest in my girlish hopes and fears.

One early December day he took me up to the Top Field to show me his horses. The elegant polo ponies were wearing smart tartan coats and after a disinterested look at Paddy and me, went back to grazing. Molly and Dolly came lumbering down to the gate, hoping for a treat and sure enough, Paddy produced a couple of old apples from his large pockets.

'Where's the pretty little pony?' I asked, not seeing her anywhere.

'Ah now, my little Merry – New Forest breed,' Paddy said, his voice softening. 'Sure, she's in her stable by my place. Tomorrow afternoon, I'm thinkin' to go into Fleet to pick up a bit o' harness from the saddlers. Come with me, girlie – do you good.'

As luck would have it, Jim had an appointment after lunch next day with Colonel Fortescue and an Agricultural official, so I was able to sneak off to Paddy's, knowing Jim didn't approve of my visits to him.

'Come and make friends with my Merry,' Paddy said, collecting up the trap harness.

But Merry did *not* want to make friends with me! Ears flattened, eyes rolling and lips curled back, she backed away from me, snorting disdainfully. I was mortified.

'Ah, she *is* Paddy's girl,' he apologised, harnessing her to the trap shafts. 'She'll get used to you.' But from the baleful looks she gave me, I had my doubts.

December had announced itself by unsettling the weather, with bouts of cold rain and boisterous winds that whipped the last of the leaves from the trees. But today was calm with a glimmer of sun and as we bowled happily along the Hampshire lanes, I felt like one of Jane Austen's young heroines setting out to buy lace and ribbons in Fleet – alas, minus the sprigged muslin dress and bobbing corkscrew curls!

We found the town quite crowded, with one or two army lorries parked and a ramshackle van right outside the saddlers, so Paddy had to stop the trap further up the street on the other side.

'You just sit tight, girlie,' he said, jumping down and patting Merry. 'She'll stand here quiet as a wee lamb, no worries,' and he set off back down the street.

Not half a minute had passed before Merry was snorting, pawing the ground and straining her neck round to see where Paddy had gone. I grabbed the reins and spoke gently to her.

'Quiet, Merry, Paddy won't be long,' but at the sound of my voice, she whinnied loudly and began prancing on her front legs, making the trap bounce up and down alarmingly. I tightened the reins and spoke more firmly.

'Stand still, Merry. Good girl. Stand still!' and I flicked the reins sharply over her back.

It was the worst thing I could have done. Tossing her head and whinnying loudly, she began backing up, the trap veering out at a sharp angle across the street.

'Oh stop, Merry, stop. At once!' I cried, really panicking now and tugging on the reins.

Several shoppers had stopped to watch. A cheeky boy on a bicycle called out 'ever 'eard of going forwards, miss,' and everybody laughed. I was *so* embarrassed.

Still moving backwards, Merry and the trap were now angled half-way across the street, me desperate, eyes searching for Paddy's return. What would he say?

It was bad luck that just then a large tractor and trailer came chugging up the street, the irate driver yelling 'get the effing hell out of my way.'

But how could I? I shrugged my shoulders, all the while Merry continued purposefully stepping back, her head craned round, looking for Paddy.

'*Go forward*, Merry!' I cried, frantically flicking the reins. 'Walk on, *PLEASE!*'.

Suddenly the wheels hit the curb behind and mounted the pavement. I grabbed the seat rail to steady myself. There was a loud grinding thud. I nearly fell out as the trap came to a juddering full-stop. Frightened, Merry reared up, whinnying, tossing her head.

I was horrified when I saw we'd backed right into a shop doorway, the trap jammed against the corner frame of two large plate glass windows. A man in the growing crowd grabbed at Merry's bridle, but she refused to be pacified, striking out with her hooves and neighing loudly. Swearing, he backed off.

A communal 'OOOOH! came from the gawping onlookers, just as the shop door opened and an extremely angry, black-suited man marched out.

'Get this contraption out of my doorway *immediately*!' he spluttered. 'You should be ashamed losing control of this animal. And the damage. I shall get in touch with my solicitor immediately!'

It was only then I saw what was inside the shop window. Headstones 'In Loving Memory', 'Rest In The Lord', with one or two ornate urns perched in folds of purple velvet. The sign over the shop read R.I. Peaceman UNDERTAKERS.

I was about to apologise profusely, but he launched into the attack again.

'What will my customers think with this commotion going on, practically *in* my premises – blocking my entrance. It's disgraceful!'

I couldn't actually see any grieving relatives queuing up outside waiting to make disposal arrangements for their dear departed, but decided this was not the best moment to point it out.

Suddenly realising Merry had quietened, I saw with huge relief Paddy holding her head, whispering soothingly to her.

'Oh Paddy,' I wailed. 'I couldn't stop her going backwards. She wanted you. I'm so, so sorry.'

'Not your fault, Jeannie,' he called, 'My Merry's been a naughty wee girlie.'

'She certainly has!' snorted the infuriated R.I. Peaceman. 'Fancy leaving this inexperienced girl in charge of such a dangerous beast! Just look at the damage to my property. *Look!*'

There were sympathetic murmurs from the onlookers as they shuffled nearer to see for themselves.

Easing Merry and the trap away from the shop window, Paddy peered hard at the so-called damage; then taking a large grubby hankie from his pocket, spat lavishly on it and gave the window frame an energetic rub.

'There now, mister,' he said, beaming at R.I. Peaceman. 'I'm thinkin' there's no harm done – not one wee scratch left.'

Everyone leaned forward to inspect. Paddy was right, not a single mark could be seen.. With murmurs of approval, the onlookers dispersed.

But R.I. Peaceman was *not* satisfied. Quivering with indignation, he shook the window frame.

'Never mind no scratches. This frame will be weakened. I demand damages!'

Still beaming, Paddy patted him on his shoulder, as if to say 'there there', and jumping up into the trap, we set off up the High Street at a spanking pace, Merry happy at last, and me, oh so relieved!

Although I was to accompany Paddy on a few more trips into Fleet in the weeks to come, he made sure Merry knew I was *not* being left in charge of her and she always stood 'quiet as a wee lamb, no worries.'

****

SEVEN

I was mostly on friendly waving terms with Bert, nosy Ethel Gummer's husband, as he navigated his tractor up and down various fields. He told me one of his other jobs was looking after the small herd of cows and heifers grazing in a far field, Uppergate pastures. The male calves, Bert said, were always sold to be reared for veal or beef.

With a stolen half-hour one morning, I walked beside Bert's tractor taking up a bale of hay for their daily feed. As we chatted, I couldn't imagine this open-faced, amiable man married to sharp, inquisitive Ethel – a puzzle.

The heifers galloped across the field to greet us, though I knew it was really cupboard-love. Their older sisters followed more slowly.

'They're mostly in-calf.' Bert told me. 'You'll be havin' 'em in to milk cum the Spring. Get some kickers among the heifers, you will. They don' like the teat-cups at first.'

I was dismayed. It hadn't occurred to me that some of the cows I was on such affectionate terms with would dry out, and nervous, newly calved heifers would come in. Just as I was feeling much more confident in the Milking shed.

I told Bert how I hankered to learn to drive a tractor, and so after he'd fed and checked the herd, he agreed to let me take the tractor back to the farm.

'You won't come to no harm,' he said. 'she is a mite past it now, but I doubt the Colonel's going t' spend out on a new one.'

How marvellous I felt, riding high, chugging across the fields in total command, even negotiating two gates before braking in the farm yard. Easy-peasy, no problem. Now I'd have another go at Jim to let me use the tractor.

An opportunity came my way soon after, when Bert had left the tractor in the field nearest the Milking shed, and walking across it, I met Jim with Masie on her way to feed the chickens.

'Not that old whine again,' Jim said, when I begged him to let me learn to drive. 'You're already driving me into an early grave.'

He wasn't pleased Bert had let me drive the tractor all the way from Upper-gate pasture – 'with *no* problems,' I pointed out, *'and I'd parked it safely in the yard'*.

'Go on, Jim,' Masie said. ' Let the girl have a go.'

'Okay, Fanny,' he relented. 'Let's see what you can do.'

I had to admit I didn't know how to start a tractor, so Jim cranked it up and got it going.

'Take her all round this field and over that bridge into the next.' he said.

My circuit round the field was easy, but I noticed the bridge had no sides, being nothing more than three wooden planks across a deep watery ditch – not much wider than the tractor. However, taking a deep breath, I approached it with calm caution. After all, it was only a matter of lining up the middle of the tractor with the centre plank, and I'd got through those two gates with no trouble. I was aware of being watched. Alf had joined Jim and Masie. I bet he hoped I'd fail.

Now the front wheels were onto the bridge, which creaked ominously. Undeterred, I kept looking ahead, holding the steering wheel steady. Large spots of sudden rain made me blink, but half way across, I began to relax. Only another two feet or so. But the left hand plank was rickety – I felt it bend slightly! The tractor tipped a little to the left. I jerked the steering wheel to the right, afraid I'd lose control. The tractor skewed over too far, so I turned the steering wheel hard left again, praying I'd straighten out.

I heard Jim shouting – I daren't think what he was calling me!

What happened next was my worst nightmare! The front left wheel slowly slid over the edge of the rickety plank. The tractor tipped alarmingly, still moving forward. I yelled out, terrified, but some instinct told me to brake. We jerked to a halt. Frozen to my seat with fear, I felt the tractor rocking gently back and forth, as if debating whether it would plunge into the watery ditch or not. By now, Jim, Masie and Alf had caught up with me.

'Bloody 'ell, Fanny! Only *you* could do this.' Jim shouted. 'Don't move a muscle; you'll upset the balance – the whole lot might go over. Now, turn the engine off.'

'Do as he says, gal,' Masie said. 'Don't be frightened.'

'I'll give her frightened!' Jim snarled. 'Alf, fetch Bert. Get some of those old posts from the workshop, and rope and the mallet. We'll have to shore up the wheel.'

Frozen fear had suddenly morphed into a strong desire to leap down and disappear; escape from what I knew might happen, but somehow I stayed put, getting soaked through, the tractor still rocking dangerously back and forth.

'I must have been mad to let you loose on a tractor,' Jim shouted, leaning hard onto the tractor back – and to my intense relief stopping it rocking.

'It's not my fault,' I wailed, 'the bridge is rotten.'

'You're a mean bugger, Jim,' Masie said, 'making her go over it. You did it on purpose – I know you!'

'Huh! Don't you interfere.' Jim snapped. 'It's Bert's fault, letting her wheedle him into driving the thing.'

It seemed like a year before Alf and Bert arrived, wet, panting and lugging several large posts, rope and a mallet.

'Here, Masie,' Jim said, 'come and lean on the back.'

'Oh thanks!' she replied, glaring at him. '*I'll* make a big difference!' Her tiny weight didn't, of course, and the tractor began to rock precariously again. I couldn't help yelping with fear. Any minute now I'd be upside down in the ditch, squashed like a fly under the tractor!

Cursing me, the rain and the muddy water, Jim climbed down into the ditch with Alf and Bert, and after several fraught attempts and lots more ripe language, they managed to construct a temporary support for the rogue wheel.

'You can shift your blinkin' arse now, Fanny,' Jim yelled.

I slid off the seat, knees weak and heart thumping. Masie gave me a big hug and I burst into tears, feeling awful I'd caused such a commotion and everyone was wet through.

'There, there,' she comforted me. 'it wasn't your fault.'

'Huh!' Jim said, glaring at me. 'Don't ever ask if you can drive this thing again; you're a menace!' and he strode off towards the dairy, leaving Alf, who was giving me dirty looks, and a rather sheepish Bert, to carefully ease the tractor over the bridge.

I didn't hang around either. Feeling silly and resentful, I rushed off to get some sympathy from Josephine. Masie was right. It hadn't been my fault. The bridge was rotten. One rickety plank had stopped me succeeding. My W.L.A. friend, Rosemary, was given six weeks tractor-driver training before she went on a farm. Why hadn't I? It wasn't fair. Josephine agreed with me, of course,

***

EIGHT

It was a good job I had no time to feel sorry for myself. Life was getting tougher now that December was treating us to bouts of cold rain and sudden, stormy gusts of wind.

There was nothing for it but to tie up my glamorous page-boy hair style in a thick scarf, forget the powder and lipstick and don my old school mack – but even that didn't keep me dry when hard rain beat down on my bent back. I'd lost my manicured nails and even creaming my hands didn't stop them getting red-raw, fingers fissured with tiny painful cracks.

Occasional wistful thoughts about that 'nice little typing job' Mother had urged me to get, floated attractively across my tired mind – hastily dismissed!

I could now just about haul a hay or straw bale onto my back, and legs buckling, struggle with it for a few yards; I could stagger around carrying two full pails of icy water with arms I swore were growing longer and coming out of their sockets, and I could now push a three-quarters full wheelbarrow of Romulus' and Remus' pungent offerings as far as the steaming dung heap. I was getting stronger, I told myself, and accepted as normal the all-enveloping ache that sent me early to bed most nights.

Colonel Fortescue often wandered about the farm, having a word with whoever was around and one day he called me over.

'Got a minute?' he asked. 'I want to show you the annexe. I've got an idea.'

I didn't know what he was talking about, but followed him to a small, flat-roofed building which jutted out at right angles to the back of the Baggs' bungalow. Unlocking a door, he ushered me in.

'I've had this designed as a small self-contained unit for two people,' he told me proudly, and showed me around the compact kitchen, bijou bathroom and large, two-bedded living room. It was quite impressive, but why had he shown it to me?

'I'm thinking of applying for another Land-girl.' he told me. 'I thought you and she could share this little place. Would you like it, eh?'

'Oh yes, I really would,' I said, thrilled to think I'd have someone to be friendly with, to confide in and share the odd grumble.

'Good good; take a bit of work off your shoulders too, yes?'

I eagerly agreed and after his usual encouraging pat on my back that always cheered me, I hurried off to the dairy. At last life was going to get better!

My happy mood wasn't to last long. Jim informed me I couldn't expect to have Christmas off. I'd have to work all through, but might get an extra hour off on the Day! For a treat there'd be no dunging out.

'The bloody cows still need milking, Fanny,' he said, seeing my dismayed face. 'They won't jump up and down and milk themselves, you fathead! You'll get a day off the week before and one early in the New Year. Think yourself lucky.'

I was devastated! Christmas Day away from my family; missing the Boxing Day party with Church Youth Club friends; it was unthinkable!

If I was miserable as Christmas approached, I noticed Monica Fortescue looking more cheerful, almost friendly. Masie told me their son, Henry, was coming home from boarding school for two weeks' holiday.

He arrived just before I had my promised day off; a silent twelve year old, whom I guessed would prefer to be back at school with his chums, rather than hovered over by doting, anxious-to-please parents. I tried chatting to him, but got the impression he considered a Land-girl beneath him – rather lower class.

With my little hoard of saved earnings, I went home to buy small gifts for Mother, my aunts and a few friends. I just had time to wrap everything up in brown paper and wool, asking Mother to put them under the Christmas tree that I wouldn't be able to decorate myself – the first time ever.

I was taken aback when Colonel and Mrs Fortescue asked me to have my Christmas Day meals with them – it would give the Larchfords' the day off from feeding me.

We were to have lunch at the Ely Hotel; a very grand-looking building near Fleet. Never having eaten in a hotel before, I was nervous I'd do the wrong thing. Having no civvies with me, I hoped my uniform would look smart enough.

Monica Fortescue was transformed! Hair groomed, make up, wearing an expensive dress and jewellery – the beautiful debutante she'd once been. I hoped Colonel Fortescue appreciated it. Conversation was stilted, with the Colonel trying to be funny, while I nervously nibbled my way through the full Christmas menu. Tea back at the cottage, after afternoon milking, was almost as nerve-wracking, the Colonel and Mrs Fortescue trying so hard to be jolly parents, as we all played one of Henry's board games. Glancing at him, I wondered if he was dreaming of midnight feasts and pillow fights in the dorm' as I was dreaming of Mother and home? Still, he was making his parents happy!

Boxing Day meant back to full farm routine and though I thought wistfully of my friends' enjoying their party, I was so pleased to relax at Masie's table enjoying Christmas left-overs and a girlish giggle with Susan and Billie.

I had to work on New Year's Eve too – another party missed, but hoped to catch up with girl friends on my day off soon after. I was feeling strangely distanced from my old life – so little time to think between endless work and much needed sleep!

I made my New Year resolutions, whispered to Josephine.

All to be broken by the end of January!

****

NINE

1946 – to be the first full year of peace since 1938. An infectious optimism was in the air, everyone looking forward to loved ones returning from the war, the gradual easing of rationing and other irksome restrictions. Would there be bananas, oranges and extra sweets, I wondered? Could I throw my clothing coupon book away and buy a new dress soon?

Dreams faded with the reality of January; cold east winds biting through my thick greatcoat, fingers scarlet and painful from breaking ice in buckets and troughs, face stiff and eyes watering. When snow fell, I didn't whoop with joy and run out to make a snowman – it was just another physical obstacle to overcome.

Alf developed a hacking cough, so Jim told him to stay at home.

'We'll have to feed the pigs, Fanny,' he announced, to my surprise.

I knew there were pigs somewhere on the farm, but didn't know Alf fed them.

'Come on, get a move on,' Jim said, looking disgruntled. 'Got to collect up all the kitchen scraps; make up a swill.'

Everyone on the farm saved all their left-over cooking waste in outside buckets, to be mixed together with bran to make the most revolting-looking mess I'd ever seen.

'Ugh!' I said. 'Poor pigs.'

'Just you wait, Fanny.' Jim laughed and thrust one of the buckets at me. 'Follow!'

'Are there any piglets?' I asked, hopefully.

'Don't be daft, this lot's for bacon and sausages – Tamworths, they are.'

'Poor things,' I said, feeling guilty I'd never given a thought to where those delicious sausages and bacon rashers came from – *when* we'd saved enough meat coupons!

The three pink pigs, I discovered, were kept in a large sty the other side of the annexe. I heard and smelt them before we even turned the corner!

'You have to watch out for the big un, Jim warned me. 'He'll knock you for six given half the chance. The smaller two are more docile.'

The excited squealing had reached a painful crescendo as Jim swung one long leg over into the sty, fending off the enormous big un' – a scary-looking monster – with his bucket and a stick.

'Move over, you great lump,' he growled, emptying the bucket contents into their trough, then grabbing my bucket to top it up. I watched in disbelief as the pigs fought each other for places to guzzle down the vomit-looking mess.

'Cleaning out every three days', Jim said. 'Think you can do it and feed them 'til Alf's better?'

I knew it was an order *not* a request, but Jim saw my apprehensive face and shook his head.

'Of all the lily-livered land-girls!' he exclaimed. 'You take the prize! You'll be alright – as long as you don't open the sty door.'

''Course I'll do it!' I said, stung to the core. 'I'm not scared of pigs!' Chin in the air, I picked up the empty buckets and strode back to the yard.

'Give those buckets a good clean out, Fanny.' Jim shouted after me.

Next morning, still chanting my mantra 'I am not scared of pigs', I stirred up the obnoxious swill and staggered with two brimming buckets to their sty. I reminded myself of my initial terror of the cows, yet now how fond I'd become of them. Perhaps, with a friendly attitude, the pigs and I could be buddies – even Big-un.

The squealing and frantic pushing seemed even greater than yesterday, Big-un standing up, his huge front hooves planted on top of the sty wall, one gigantic mass of quivering pink flesh – taller than me! The smell was almost over-powering, my courage trickling away fast as I worked out how best to cope with the tricky situation.

Maybe if I spoke gently he'd react better than to Jim's stick and gruff 'move over, you great lump'. After all, the cows had quickly responded to me – even Primrose had stopped kicking me.

'Hello little piggy, nice piggy,' I cooed, 'Get down now – *manners*!'

He obviously couldn't hear me over the noisy commotion they were making, so I tried again, louder.

'NOW BE A GOOD PIGGY, BIG-UN. GET DOWN, THERE'S A DEAR AND YOU CAN HAVE YOUR DINNER.'

He understood the word 'DINNER' because he got more excited than ever; barging up and down, treading on his smaller sty mates, his beady eyes fixed greedily on my buckets.

With my last vestige of courage and protecting myself with a bucket, I quickly flung one leg over the sty wall – but Big-un was even quicker. Before I'd had time to lift the other leg over he'd charged the bucket, slopping half the messy contents down my dungarees and the rest over himself, pinning me hard against the wall. With delighted squeals the other pigs joined in, dragging the bucket down to the ground and greedily devouring every last scrap, while I gasped with pain, desperately wondering how to extricate myself.

I was being split in two astride the wall, my outside leg getting excruciating cramp and the other being agonizingly pummelled by Big-un and Co. It seemed forever before I could find the right moment to quickly retrieve my tortured leg, nearly falling backwards and almost spilling the rest of their dinner.

No way was I going to climb in with those demented pigs! Giving a quick glance round to see if anyone was watching, I flung the swill in the general direction of the trough – just as Big-un hurtled towards me. Oh horror! Very little swill reached the trough – most was sliding off Big-un's snout and ears, the rest cascading down the other pigs' backs. What a ghastly mess! I was going to be in major trouble; again!

But to Big-un and Co. food was food wherever it landed and to my fascinated relief, they sucked and slurped up every last morsel off each other, eventually revealing three remarkably clean pink pigs and even a passable sty!

As I limped back to the yard, reeking of pigs' swill and dung, certain I was maimed for life and dreading the thought of cleaning the pigs out tomorrow, I bumped into Jim.

'What's up with you, Fanny?' he asked, looking at my pummelled leg and my battle-scarred dungarees.

'HORRIBLE PIGS! BLASTED BIG-UN!' I shouted, not caring any more how many sausages they made out of him.

'Give you a hard time, did he?' he said, sounding almost sympathetic. 'Okay, I'll do 'em tomorrow. Go off home for dinner. Get Masie to look at that leg.'

It was already getting colourful as Masie rubbed some cream into my bruises, insisting to Jim I rested until milking time.

Alf came back to work next day and luckily I was never asked to feed Big-un and Co. again.

Not only was my leg badly bruised, but ditto my ego. It was depressing to accept I had no rapport with belligerent bulls, a mean-eyed horse and now pigs – especially Big-un. Even the tractor had ganged up on me!

Once again, to cheer myself up, I sought consolation with Josephine and her gentle sisters.

****

TEN

When I heard Masie say to Jim, 'It couldn't be a worse week for the Withers brothers', I didn't know who or what she was talking about, except we were now getting really stormy weather; even more uncomfortable to work in than snow!

Jim said the Withers were hedgers and ditchers, long established, who travelled from farm to farm throughout Hampshire to prune and repair hedges and clean out ditches – a skilled craft, he said – something I'd never realised.

I met them quite soon when Masie asked me to take out a jug of tea to Wilf and Ern at dinner times each day.

I found them at the far end of the lane, already sitting on the grass verge, eating door-step sized sandwiches.

They stared so hard at me, I thought perhaps I had mud on my face, or my nose needed blowing.

'You'm one of they land-gurls!' the gaunt, very thin one exclaimed, unwinding himself into a tall willowy bean-stalk.

'That she be, Wilf, she be one,' his plumper brother, Ern agreed, chomping on his sandwich with what looked like toothless gums.

Feeling uncomfortable, I told them my name and put the jug down, hoping they'd decide I was just an ordinary girl, not an oddity to be stared at.

'Thank 'e, miss,' said Wilf, not taking his eyes off me. 'We comes across lots like you, don't we, Ern?'

'We does, Wilf, we does,' he agreed between chomps, nodding vigorously like a demented mandarin.

I stood there, wondering just where this conversation was going, and if 'coming across lots like me' had been a good or a bad thing?

'Are there any land-girls nearby?' I asked, thinking how nice it would be to meet them.

Wilf and Ern looked at each other for a long time, frowning.

'None hereabouts we knows of, are there Ern?' asked Wilf.

'None hereabouts, Wilf, none,' he agreed, shaking his head furiously.

They continued to stare, so waving goodbye, I hurried back to my bedroom to inspect myself in the mirror for any peculiarities I may have developed – but couldn't find a single one.

Over the next week, try as I might, conversation with the Withers never progressed much beyond acknowledging I was 'one of they land-gurls', and there were 'none hereabouts' – with repeated shakings of heads.

They did appreciate my admiring the tidy transformation they brought about to the farm's hedges and ditches, exchanging proud, satisfied nods.

'We sees lots like you, miss,' Wilf finally said, shaking my hand with hard bony fingers, 'but you's a sight for sore eyes, ain't she, Ern?'

'For sore eyes, Wilf, a sight she be.' he agreed, his head nodding so much I was sure it would fall off!

Buoyed up by their compliments, I was unprepared for Jim's bad mood in the milking shed that afternoon. Perhaps he'd had a row with Masie? Try as I might, I couldn't do a thing right; too slow, too frantic, too careless – my faults were endless.

With milking at last over, my arms plunged deep into a steaming tank of dirty utensils, I allowed myself a good cry. Tears part-anger, part-misery, cascaded down my cheeks onto my scrubbing brush and into the water.

Suddenly, to my embarrassment, I found Colonel Fortescue standing beside me.

'Bit hard on you is he, eh?' he said gently.

I nodded, grateful for his sympathetic tone.

'Cheer up m'dear,' he said, putting an arm round my shoulders. 'Don't want to take too much notice of Larchford. Bark's worse than his bite, d'you know.'

'You're very kind,' I said, blinking up at him. 'I thought he'd ask you to sack me.'

'Silly little gel,' he laughed, squeezing my shoulders. 'Now, I've got a splendid idea, buck you up no end. M'wife's away at her sister's for a day or two. Why don't I fix us a nice supper tonight; have a lazy evening by the fire, eh – would you like that?'

'Oh yes, thank you, I would.' I nearly burst into tears again, he reminded me so much of my adored, long-absent Father.

Colonel Fortescue pressed a large handkerchief into my hand, patted me on the bottom and was gone.

I managed to dry my tears before telling Jim, in a stiff voice, that I wouldn't need supper tonight and would he tell Masie.

'Oh, so where are you going?' he asked suspiciously, but I pretended not to hear and disappeared as soon as all the jobs were done.

As I eased off my muddy gumboots in the cottage scullery, an appetizing smell nosed out from the kitchen. Over a noisy clatter of pans, Colonel Fortescue called out.

'Not quite ready yet, m'dear. Why not go up and have your bath, eh. Don't bother to dress again – waste of time, right!'

Happy to agree, I luxuriated in a hot bath and anticipated a delicious supper cooked by my very kind employer.

Scrubbed and glowing, clad in my sensible striped-winceyette pyjamas, I appeared downstairs to find the small sitting room table laid for two, and drawn up before a roaring log fire.

A smiling, velvet-jacketed Colonel Fortescue stood holding a bottle and two glasses.

'Down the hatch!' he said heartily, handing me a glass.

'What is it?' I asked, suspiciously. 'I don't drink.'

'Time you started then,' he replied. 'It's only wine – won't hurt you.'

I took a sip, shuddered and pulled a face. 'I, I'm sorry,' I said, embarrassed. 'I don't like it.'

'Funny little thing! Never mind.' Colonel Fortescue led me to the table. 'Try the chef's speciality – French Onion soup.'

Very hungry, I tucked into everything he put in front of me, praising each course with grateful enthusiasm. In and out of the kitchen he darted, beaming down at me, tossing back glass after glass of wine.

'Got to drink for two now, haven't I, eh!'

The meal over, he pushed back the table and pulled the sofa up to the fire.

'Leave the crocks, m'dear. Come and sit here.' He patted the sofa by his side. 'How about reading to me? Jolly fond of it, d'you know.'

I hadn't bargained for this. I felt too shy to read aloud, but he put a pile of magazines on my lap, and said, 'Read anything, I'm not fussy.'

Flicking quickly through a copy of Punch, I began to read at random, stumbling nervously over the long words, apologising at every other sentence.

Suddenly I became aware of Colonel Fortescue's arm sliding round my back. His fingers began inching up inside my pyjama jacket.

Puzzled, I asked, 'What are you doing?' He obviously wasn't paying much attention to what I was reading.

He brought his face close to mine and whispered 'Counting your ribs, you quaint child.'

'Are you interested in anatomy?' I asked eagerly.

'Umm, you could say that,' he chuckled and nibbled my ear. Extraordinary!

His hand tightened round me, fingers climbing up even higher until they were just touching my left breast.

'Oh, you needn't bother to count my ribs,' I told him, moving his hand away. '*I* can tell you.'

I hadn't come top in biology at school for nothing and was especially good at naming all the human bones.

'We've got twelve pairs of ribs,' I informed him, 'the bottom pair floating, meaning they aren't attached to the sternum, the breast bone that is.'

Colonel Fortescue was looking quietly impressed.

'If you're interested,' I said, 'I've got a jolly good book upstairs showing the skeletal structures of all mammals.' I'd brought it with me so I could learn how to draw farm animals.

'I'll fetch it.' I told him.

'Don't bother,' he said, jumping up and beginning to clear the table.

'It's no trouble,' I assured him, and in another minute was downstairs again carrying my large illustrated biology textbook.

I opened it on a double page showing detailed drawings of male and female human skeletons and called him over to the table, but he was throwing cutlery into the sideboard drawer and to my great disappointment seemed to have lost interest entirely.

'Better get off to bed,' he muttered, sounding rather grumpy I thought, as he stomped noisily around the kitchen.

Laying in bed that night, I puzzled about how unpredictable some people – and animals – could be. You never knew how they'd act, so you didn't know where you were with them?.

It was *very* bewildering!

****

ELEVEN

Anxious to know when the new land-girl was arriving, I asked Jim.

'Gawd forbid, Fanny! Aren't I plagued enough with *one*? Where'd you get that idea?'

I repeated what Colonel Fortescue had told me and how the two of us would share the annexe.

'Huh, he'd like that,' Jim said. 'Not a hope in hell! You girls are in short supply I've heard – thank the Lord. Two of you doesn't bear thinking about.' But seeing my furious face, he laughed.

'Only joking, Fanny. Anyhow, the annexe is for the new gardener.'

'Gardener!' I exclaimed, yet another dream shattered. 'Why?' But then remembered Mr Attlee was again urging British people to grow more food. It was in shorter supply than ever, our rations cut again. Jim said a big piece of field near the annexe was to be dug up for growing vegetables and fruit. The new gardener, Roy Pengelly, his wife and baby would be arriving soon.

Ethel Gummer was none too pleased. She didn't want 'babies crying all night under my bedroom window'. I almost said 'it's only *one* baby'. She was none too pleased with me either for bringing her a load of extra filthy dungarees and socks and made sure I paid her extra.

Looking forward to another longed-for Sunday off, I was upset to get Mother's letter asking me not to come home. My aunts complained I smelt of cows and as the Vicar was coming to tea, it might offend him. Very indignant, I nearly wrote back to remind them that Jesus was born in a cow shed and I bet his swaddling clothes smelt a bit, so the vicar shouldn't mind! Then I recalled my aunts throwing windows open every time I went home so guessed they didn't appreciate my healthy aroma.

Just before the Pengellys' arrived, the skies darkened and February launched a very wet spell, making it hard work for ever-cheerful Bert to plough up the garden area ready for planting.

Wondering what to do on my Sunday off on the farm, the sun suddenly put in a fleeting appearance, so I took my sketch book outside and tried drawing Josephine, the chickens, who scratched around much too quickly, and even mean-eyed Merry glaring at me over her stall door. More practise was definitely needed, with closer reference to my anatomical biology book.

I was busy milking with Alf when the Pengellys' moved in so missed the excitement, but was sure Ethel and Doris had noses glued to kitchen windows. I had to wait until Masie came over with our tea before I got any details.

'Well, gal, I haven't seen her yet,' she confided, 'but if he's a gardener, I'm a Chinaman! Thin as a reed – I could pick him up, easy. Wherever did the Colonel get him from? Beats me.'

Jim called by and said exactly the same, only in more colourful words, so I could hardly wait to see this weedy human specimen.

I just hoped Mrs. Pengelly wasn't another Ethel or Doris and I wouldn't be asked to kiss a snotty-nosed baby.

As it was, I was the first to see her the following morning after milking. Passing near the annexe, the door suddenly opened and out came a pram followed by – I caught my breath – *the most stunning vision* I had ever seen! Burnished red hair cascaded abundantly around a creamy, flawless face and down over a taffeta orange-and-black dressing gown. A Pre-Raphaelite beauty – Dante Rossette's muse reincarnated!

She looked up with huge brown eyes and her wide, cupid-bowed mouth smiled at me.

'Come and say hello,' she called, her voice soft and musical.

I realised I was staring open-mouthed. I just couldn't help it.

'Eh, how do you do,' I said, pulling myself together. 'I'm Jeanne.'

'Flora,' she told me – how could it be anything ordinary?

I felt obliged to peep into the pram, hoping desperately the baby didn't take after Dad and I'd have to admire it, but the pretty, sleeping face on the frilly pillow was a miniature copy of Flora.

'She's sweet!' I said, meaning it.

'We've called her Lily.'

It was hard to drag myself away, but the yard had to be swept. I'd just have to wait to meet Flora's weedy husband.

I was realising I still had lots to learn about my cows. Some were slowly producing smaller milk yields and would soon dry off completely at about seven months in-calf. Jim told me these cows were especially susceptible to swollen or tender udder quarters, and we had to watch them carefully. If a teat produced stringy or lumpy, discoloured milk, the usual treatment was gently rubbing the quarter and stripping the teat out until it was cured. He said if a cow developed more serious Milk Fever, the vet, a Mr. Blunt, was called and an immediate injection of calcium cured it quickly. The worst they could get was Mastitis, a very serious udder inflammation, but luckily none of the cows developed it while I worked there. By law, the vet came twice a year to check the herd was tuberculosis free. I'd met Mr. Blunt once, a large, big-voiced man with a brusque manner, when he'd paid an Autumn visit to stitch up Bonny's teat she'd torn on some brambles. She didn't appreciate his help and I needed all my strength to hang on to a back leg to prevent Mr. Blunt having his head kicked in.

The weather having turned suddenly mild and sunny meant Bert finished ploughing the garden area quickly and was just fixing the harrow to the tractor when I spotted the elusive Roy Pengelly talking to him. It was true; he was a small wisp of a man who looked as if the slightest breeze would whirl him away.

Whatever did the gorgeous Flora see in him?

The Pengelly's arrival had certainly caused quite an stir amongst the quiet, routine-programmed inhabitants of Brinkleigh Farm, as if the atmosphere had been ratcheted up a notch or two. Nobody was acting the same, I noticed; opinions flew about freely and I was often the eavesdropper. Ethel and Doris went around with permanently disapprovingly, pursed-up lips, scandalized that 'she goes *outside* in that flimsy dressing gown with the *low* front!'

Bert had an extra big grin on his face and proclaimed Flora was a 'corker'. Alf just looked dazed, his shiny round face rosier than ever and his little current eyes nearly popping out of his head.

To Masie, Jim was foolish enough to say, 'Cor, if I was ten years younger, I'd be in there giving her a helping hand!'

'You just keep yer hands to yourself and your eyes off her,' Masie told him, poking him in the chest, 'or your life won't be worth living!.'

'She means it,' he told me, nodding. '*Oh* yes!'

Colonel Fortescue seemed to be paying the Pengellys a lot of visits, and Flora was getting the pats on the back that he no longer gave me – in fact, he totally ignored me these days. I fathomed he hadn't liked me knowing so much about ribs; men always like to know best!

Monica Fortescue walked about with her usual inscrutable face and I guessed she was used to her husband patting young women.

I'd seen Paddy knocking on the Annexe door a couple of times too and when I next went to visit him, he was tossing back a *large* somethin' in a glass, 'to toast the little darlin' beauty that's come into our midst'. He told me he'd offered to take her into Fleet any time she wanted and promptly poured himself another large somethin' for a second, or was it a third toast?

Even I felt different. I looked at myself critically in the bedroom mirror. No creamy, flawless face; just a blotched, weathered one with chapped lips; dry brown hair wisping out of an old head scarf and rough-red hands I tried to hide when I went home. Inspecting my body, I saw no 'Flora' curves – I'd gone quite scrawny despite Masie's good cooking. Flora had *no* competition at Brinkleigh farm.

There was only distain showered on poor Roy Pengelly, but as the days went by and he was seen to tackle the garden project with efficient energy, the criticisms petered out.

'Well, I'll be damned; he knows what he's about,' Jim pronounced. 'Blowed if he isn't strong as a little ox despite he's like a blade of grass.'

It was several days later when I took my dirty washing to Ethel that I heard her puzzling reason *why* he was like a blade of grass. I knocked the open kitchen door and hearing voices, peered round it to see Ethel and Doris with heads together, deep in a conversation I couldn't help overhearing.

'– and in daytime too,' Ethel hissed. 'I often sees him popping back. And what for, I ask myself? Not for a rest, I can tell you.'

'Well I never!,' Doris gasped.

'I blame *her*,' Ethel whispered. 'she's one of them women what tempts men so they can't help themselves!'

'Have you heard them – *doing* it?' Doris asked.

' Huh, not just heard. I *seen* 'em, haven't I!'

'Ooh!' Doris breathed. 'How?'

'Well, if I stands on my bed,' Ethel confided, her nose nearly touching Doris's, 'I can just see their bed out of me window. *At it*, they are, mornin', noon and night – *at it* like rabbits!'

'*Disgusting*, I say!'

'Poor man, no wonder he's so thin.' Ethel said indignantly. 'She's sappin' his strength, you can tell.'

So intrigued was I with this fascinating conversation, I'd forgotten they didn't know I was there, so I went outside again, and looking innocent, knocked the door really loudly.

Ethel was obviously irritated I'd disturbed them and she got rid of me very quickly, while I went away really puzzled by what she and Doris had said.

What was the *'it'* Flora and Roy were *'at'* that made him so very thin? Why did it upset Ethel and Doris so much – and what had *rabbits* got to do with it?

Maybe I'd ask Masie – she was bound to know.

* * * *

TWELVE

All negative comparisons between myself and Rossetti's muse, the beauteous Flora, (*who'd reduced her husband to a mere blade of grass in some mysterious way*), were swept out of my mind by Jim's enthusiastic spring-cleaning projects.

'Before we know it, Fanny,' he announced, 'it'll be carving time. Got to get the stalls ready and put a lick of whitewash on the shed walls. Nice job for you – fill in your spare time. I've noticed you skiving off to see that Irish layabout, and disappearing to powder your nose, whatever.'

I opened my mouth to protest but he carried on.

'Stone the crows, Fanny, you're here to work – not knitting for the troops. You've had it easy so far. Wait 'til Spring comes; you won't know what's hit you.'

I was too flabbergasted to argue. '*Slave-driver*', I muttered, glaring at him; wondering how I could squeeze out any extra energy to cope with even more work.

Next morning, immediately after milking he marched me off to a long shuttered shed I hadn't been in before. Down one side was a line of about eight empty stalls, bereft of straw; the whole place smelling musty, with cobwebs festooning the rafters and the floor crunchy with dead insects.

'First job, we throw open the shutters,' Jim said, 'Get your skates on! Let's have a good blow through.'

We hadn't even finished doing that before Jim was pointing out a large broom and long handled brush propped by the door.

'Get all those cobwebs down and give the floor a thorough sweep, Fanny. Half an hour's work, then you can have your breakfast. I'm off to take the cows back.'

Feeling murderous, my hollow stomach in revolt and a bone-chilling, Siberian wind tunnelling through the open windows, I took nearly an hour to de-cobweb and sweep up the corpses. Dusting myself off, I staggered back to the Larchford's bungalow, expecting to find I'd missed breakfast, but kind Masie had saved it. If it hadn't been for her, I might have packed my bags and left that day!

Before afternoon milking, Jim set me up hosing down the shed and sweeping the dirty water into a central gully. No problem with drying; Siberia was sending over even stronger, face-numbing winds and Bert was studying the sky, shaking his head and forecasting snow 'afore too long'.

I was allowed my breakfast straight after milking the following morning.

'Masie's had a go at me,' Jim grumbled, sounding indignant. 'We're not finished yet, Fanny. I've mixed up the whitewash and you can give me a hand slapping it on the walls, two coats. When it's dry, I'll get Bert to bring along some straw bales and you can bed-up the stalls. A nice easy task, eh?'

With not a minute to recover between milkings, dunging out, lugging straw, replenishing water buckets, walking Romulus (hardly ever Remus), and gobbling down meals, I worked non-stop, trying to keep up with Jim as we freshened up the shed walls.

Wielding the large, whitewash laden brush, my arm began to feel like lead, my neck got horribly cricked and I lost all feeling in my legs! To add to my misery, the relentless wind whistling through now carried driving snow and I soon resembled a speckled snowman!

Left to bed-up the stalls, my body complaining bitterly as I staggered about with heavy fork loads of straw, I heard Mother's pleading voice again 'get a nice little typing job down the road, dear'. If only I'd listened! But then, how I'd miss my girls, especially Josephine, and as I stood back and admired the dazzling white walls and the waiting stalls bedded up with sweet-smelling straw, I had a real sense of achievement.

Hadn't I joined up to serve my Country in her Hour of Need? I knew the answer.

What I didn't know until my left sock began to feel wet, was that I'd stuck the pitchfork through one gumboot. Disaster! The snow turned to slosh, the ground to mud and hopes of a new pair were dashed by Mrs Bartholomew. There could be no replacements for some months; rubber was still in frighteningly short supply.

So when Jim sent me out to hoe a big field of turnips, my left sock was soggy, my foot blue with cold, and at the sight of those rows and rows of turnips stretching endlessly ahead, my spirits descended to join my wet, frozen toes. Half bending to hoe the heavy earth sent tight spasms of pain around my already aching back, and despite my thick overcoat, the wind cut right through me.

The next morning I could only roll out of bed and when I arrived for milking, one foot would hardly follow the other, but knowing I'd get no sympathy from Jim, I gritted my teeth and carried on.

I'd just let Bella into her stall and bent down to wash her udder, when a sudden, searing band of pain shot round my lower back, making me cry out and collapse in a heap under Bella.

'What the devil's up with you?' Jim asked, looking down at me.

'My back's gone!' I groaned.

'Nonsense!' Jim said, grabbing my arm to pull me up; but I shrieked with pain again.

'Don't touch me,' I begged. 'I *can't* move.'

'And you can't stay there for ever, Fanny,' Jim said. 'Got cows to milk.'

'Leave me alone,' I moaned, 'You get on with them.'

With a shrug, Jim carried on, and apart from the occasional look at me, milked the rest of the cows, while Bella 'mooo-od' and fidgeted, stamping her hooves and flicking her tail impatiently, anxious to have her full udder emptied and get rid of this strange, lumpen heap lying under her back legs, uttering peculiar mewling noises.

'I'm going to get Bert to give me a hand getting you up,' Jim decided, at last taking me seriously. 'Got to get Bella milked.'

Together, they slowly lifted me into a painful 'S' shape. I panicked I'd never be able to straighten up, let alone ever walk again.

'We'll have to get her to the doc',' Bert said, so Jim went across to the Fortescues' cottage to see if he could borrow their car.

I was surprised when Mrs Fortescue came hurrying over, and taking one look at me, volunteered to drive me to the doctor's. With great difficulty and much yelping from me, Jim and Bert managed to hoist me onto the back seat where I huddled in total misery.

The doctor had to get his nurse to help lever me out of the car and lift me, still 'S' shaped, onto his examination couch.

'Well, young lady, what seems to be the trouble?' he asked – the daftest question! Wasn't it obvious?

I pointed feebly to my back and as Mrs. Fortescue explained to him, I thought how like our vet, Mr. Blunt, the doctor looked; large, big-voiced and brusque.

'Soon have you back on your feet again,' he boomed cheerfully, laying a heavy hand on me. 'Let's get you flat on your tummy; come on nurse,' and before I could yell '*OH NO!*', they'd straightened me out, despite my anguished wails.

'I'll give her a shot of pain relief,' the doctor told Mrs Fortescue, while the nurse tugged off my dungarees and pants.

'How long before it takes effect?' Mrs Fortescue asked.

'Umm, about an hour,' he said brightly, 'then she can go back to work.'

I couldn't believe my ears! I thought he'd say 'give her the week off, poor little lass,' and I could crawl home to be cosseted by my doting Mother.

I couldn't believe my eyes either, when I saw him approaching me with a vet-sized syringe, sporting what looked like the longest, thickest needle I'd ever seen.

'Keep still now,' he told me, 'this is going to hurt!' and he slowly inserted the needle into my lower spine, while I clamped my teeth together and quietly moaned, trying not to yell out despite the excruciating pain.

'Not too bad, was it,' he told me, as if daring me to say *'it bloody well was!'*, then giving my bottom a much too vigorous rub.

'Thank you, Dr. Blunt,' Mrs Fortescue said, shaking his hand.

He *was* the vet! I'd guessed right. Or was it his twin brother? Whatever – both were Blunt by name and by nature.

Still unable to stand up properly, I was scooped back into the car and deposited at the Larchfords. By now, the clamped sensation round my back, like an over-tight elastic band, was easing and after an hour the pain had miraculously gone!

Despite Masie's pleading, Jim soon had me back on his idea of 'light duties'.

Not even the day off!

* * * *

THIRTEEN

February slipped into March without anyone saying 'Ah, spring is on the way.' There were no hares boxing, buds bursting, birds singing and building nests, or primroses opening their petals to the sun. Only small clumps of snowdrops braved the relentless cold that held us, all hunched up, endlessly in winter. Even the expectant mothers decided not to drop their vulnerable calves, despite Jim watching vigilantly for early signs, with every care and comfort at the ready.

In the lull, with no extra jobs Jim could nag me into doing, he gave me two days off. I gratefully hurried home, desperately needing to buy several extra pairs of thick socks to relieve the misery of a constantly wet, frozen foot. Jim had tried to repair my holey boot with a bicycle puncture patch, but it fell off almost immediately.

At home, Mother's pampering, which I usually lapped up, became indignant concern, chorused by my three aunts. Trying on a row of proffered household gumboots, I felt like Cinderella – but none of them were big enough. I was made to promise to write to Mrs Bartholomew and insist on new boots. The aunts joined my worried Mother in declaring me 'looking neglected and over-tired.' Not what I wanted to hear when I'd just been to the cinema with a friend to see glamorous 'blonde-bombshell' Betty Grable fluttering her long lashes at an attentive hero.

Arriving back at the farm two days later just in time for early milking, I realised something was wrong. The cows were still 'in bed' and nobody about, though the vet's car was parked outside the cottage. Then I spotted Masie running towards me.

'Jim says to get on milking,' she called, 'Alf's just coming,'
'What's up?' I asked.

'All hell's broken loose,' she told me. 'Can't tell you now; got to get the kids ready for school.'

I hoped Alf wouldn't be long. I'd never done the milking single-handed before, but just as I was letting the first six cows in he turned up, looking more po-faced than ever. I asked him what had happened.

'T'aint my business to say,' he muttered, 'All I knows is it wasn't nothin' to do with me.'

And when I did finally find out, shocking though it was, I was very relieved that it was nothing to do with me either. Jim hadn't turned up for breakfast, so Masie told me the whole sorry story.

The morning after I'd left for home, Bert spotted one of the Colonel's polo ponies laying very still on its side. Digging out Paddy – who was probably just raising his first glass of a little somethin' to toast beautiful Flora – they discovered with dismay the animal was dead. There was nothing for it but to tell the Colonel, who after his first angry disbelief, telephoned the vet, Mr Blunt. He arrived soon after and having called Jim, they all hurried to the far field where the horses were. Examining the pony, Mr Blunt couldn't find any obvious cause for its death, though post-mortem tests would be done. After checking the other ponies and carthorses, Molly and Dolly, he did say they all looked in need of better care.

'The Colonel was furious,' Masie continued, 'Didn't just blame Paddy, but Bert and Jim got it in the neck too. Bert takes the hay and straw up, he said, and it was Jim's job to keep an eye – he's in charge.'

For the first time ever, I felt a bit sorry for Jim. Even I realised Paddy was taking too many wee rests after too many wee toasts, but couldn't believe he'd neglect his beloved horses.

'Oh, there's worse to come,' Masie carried on breathlessly. 'They took the dead pony away. Then yesterday morning – you'll *never* believe it – the other two were found up there, dead as doornails.'

I couldn't believe it, as Masie went on to tell me the Colonel had nearly exploded with anger, sacking Paddy for gross neglect, haranguing Jim, Bert, even Alf, and anyone else he came across, including Roy.

'He's got them all in the cottage now with the vet – a pep talk, I expect.' she said. 'Jim won't be fit to live with. You'd better watch out too, girl.'

I fully intended concentrating one hundred per cent on doing everything right so I didn't incur Jim's wrath, and to keep out of the Colonel's way.

Much later that day I sneaked over to see Paddy – I wanted to say goodbye – but he'd gone, the place empty, though still full of his warm, whisky-perfumed presence. Even Merry and the trap were gone. I felt bereft, already missing a true friend. I imagined him now, the trap piled high with boxes and sacks, gently urging Merry on in his musical voice, as she pulled him further and further away from Brinkleigh farm into an unknown future. I wondered where he'd go and what would become of him? Perhaps he'd go home to find his wife and son? I was never to find out.

I was never to understand exactly what the ponies died of either and couldn't really accept it was all Paddy's fault, whatever the others said!

*** *

FOURTEEN

One permanently wet foot and my bathroom towel rail festooned with steaming wool socks had become just another discomfort to be endured, while the skies sent down alternate snow or rain showers and the east wind bit through however many layers of clothing I put on.

My plea for new boots had been ignored, but suddenly a large parcel arrived – not only boots, but two pairs of socks. To add to the bliss of warm dry feet, the wind suddenly dropped, the sun shone and Spring hurried to catch up.

So did the expectant mothers; calves began to arrive and I witnessed, quite astonished, the miracle of birth, occasionally and fearfully helping Jim when a cow had difficulties – the calf not presenting in the normal way – by gently tugging on spindly legs, while with his arm right up inside the birth canal, Jim sorted the problem out.

The stalls began to fill up with proud new mothers and adorably pretty calves, but I was shocked to realise they would be separated after a few days, the cows to join the milking herd; the calves to be hand-reared. The upsetting sound of mothers' crying for their offspring went on for several days and nights.

I so enjoyed helping Masie feed the calves twice a day. She showed me how to hold my fingers up in the bottom of a pail of skimmed milk – emulating the mother's teats – so that the calf could suck up the milk. It didn't take them long to learn.

'Only skimmed milk?' I asked Maise, but she told me that although the cream was taken off first – and sent over for the Fortescues' use – the essential nutrients were still left in.

I knew the male calves would be sold off quite soon, so we always hoped for a lot of heifers.

I didn't enjoy milking first-calf heifers; wasn't prepared for their nervous reaction to the milking shed, its noises and new sensation of having teat-cups fastened onto sensitive teats – which they immediately tried to kick off, sometimes succeeding – but usually catching me instead! Soothing talk didn't calm them and I soon displayed a painful array of multi-coloured bruises, though the worst weren't in sight! That Jim didn't get kicked much was very annoying and if he did, he never bruised. Thick skinned, I decided, and no feelings.

He had been right, I was busier than I'd ever been, but the continuing hot sunshine not only de-froze Mother Earth, it warmed my aching body and lifted my spirits, so I didn't mind as much leaping out of bed at four forty-five every morning and faced each day happily with new energy. There were no days off while we were so busy.

April arrived still very warm and fine, and while I was revelling in the heady cocktail of spring sights, sounds and perfumes, Jim had his eye on the burgeoning greens, gauging the time we could let the cows out to pasture. They were restless, I could tell, eager to taste the delicate new grass they could already smell. But just before that day came, Bert's trailer carried up a great bundle of posts and wire fencing to Long meadow.

'Can't let the girls eat too much new grass, Fanny,' Jim told me, 'not good – blow 'em out. We divide the field up.'

I helped him unroll the bundle and drive in the posts across the first third of the field. Only then did I realise it was an electric fence, to be respected not only by cows, but humans! I soon learned how careful I'd got to be when Jim tested the current and I stupidly touched the pulsating wire. Never again. The cows learned fast too!

It was a joy to see them eagerly galloping up the lane to Long Meadow, Alice and Princess jostling to be first and even old Duchess breaking into a sedate trot, her low-slung udder swinging wildly from side to side, while I skipped merrily along behind.

With my girls' thoughts happily fixated on succulent grass, mine turned – unexpectedly – to *romance*! Well, it was Spring!

As I walked the cows up and back to Long Meadow in the following days, I could see a tractor busy in a field belonging to the next farm, just the other side of the lane.

As it worked its way closer to the hedge, I spotted a young man driving it, and an even closer inspection revealed a divine Adonis, bronzed and blonde haired, smiling down at me, making my heart skip a beat! I shyly smiled back, relieved I'd washed my hair the night before and wasn't wearing a ghastly scarf.

Quite casually, I asked Jim about the farm.

'Belongs to Samson, arable farmer, quite wealthy. A big set-up,' Jim told me, 'Got three sons – two still away in the R.A.F; youngest helps his Dad. Why d'you want to know?'

'No reason, just interested.'

The next morning the Adonis smiled and waved at me, and quite overcome, I waved back, glad I'd put some lipstick on. Disappointedly, he didn't drive by when I took the cows back, but after evening milking he was there again and stopping the tractor, threw a piece of paper at me across the hedge and then drove off. Surprised, I scooped it up, but had to wait to read it until after my girls were safely in their field.

'MEET ME BY THE OAK TREE TOMORROW AT 7' it said in big capitals. Oh, how romantic!

My first date *ever* – and with a wealthy farmer's son! What would I talk about? How should I behave? I was nervous but excited and asked Masie if I ought to go.

'Course you should, gal,' she said, 'time you had a bit of fun.'

My lack of concentration next day was quickly spotted by Jim's eagle eyes and while I was at the big sink scrubbing the utensils after milking, he came up behind me, and much to my surprise, put his hands round my breasts.

'Now, Fanny,' he said, cupping them firmly. 'I'm pretty sure you don't know a thing about the birds and the bees, eh? You want to be careful going out with that fella'. Don't let him to do this,' and he gave my breasts a couple of squeezes.

'Why would I?' I said impatiently, pushing his hands away.

'Cos' that's the way you get a bun in the oven, see? Don't want you joining the puddin' club, do we?'

Buns? Puddings? Birds and bees? Whatever rubbish was he talking? I just shrugged and got on with my scrubbing.

'Just you get back before dark too,' he called as he left. 'Got to get up for milking next morning.'

What a fuss, we were only going for a short walk. After all, we hadn't even been introduced!

My date was already standing by the oak tree when I arrived, the evening sun shining like a halo on his blonde curls, a big smile on his handsome, tanned face. I was struck dumb with admiration!

' 'ello, you came then,' he greeted me.

Did I detect a missing 'h'? Not knowing what to say first, I stuck out my hand.

'How do you do,' I said, 'I'm Jeanne Parsons.'

He laughed. 'Oah – 'ow d'you do yourself,' and he pumped my hand vigorously. 'M'name's 'erbie.'

Although dismayed by those missing 'h's, I nevertheless persevered.

'It's a lovely evening.' I ventured. 'Shall we go for a walk?' and to cover my embarrassment, I set off up the lane at a fast pace.

'Oy, wait for me,' he said, and put his arm round my waist.

Embarrassed, I pulled myself away and began walking faster.

'What's your 'urry?' he asked, catching up and grabbing me round the waist again.

'Don't do that, please!' I told him, trying to shake his arm off.

"Ere, I thought we was on a date?"' he said.

'We are, but we've only just met,' I explained.

'You pullin' me leg?' he asked, laughing. 'You're a rum one!' and he pinched my bottom.

Shocked, not knowing what to do or say next, I thought I'd better make a concession.

'You can hold my hand if you like.' I said, feeling generous.

'Oh *thanks*, Miss La-di-da, 'ow kind of you,' he sniggered, and grabbing my bottom tightly in his huge hand he squeezed it hard.

Panicking, I could only think to walk faster and faster, trying unsuccessfully to shake him off, until we came to the end of the lane and I saw a chance to escape.

'Must go home now; got to get up early for milking,' I told him. "bye,' and wriggling free, I fled back towards the farm.

'AINT NEVER MET A PRISSY LAND-GIRL LIKE YOU BEFORE,' he shouted, 'DON'T WANT TO NEITHER.'

Highly indignant and bitterly disappointed, I hoped I'd never meet anyone like *him* again. My handsome, bronzed Adonis had crashed off his plinth. I could have forgiven the dropped 'h's' but *not* the shocking over-familiarity!

I certainly wasn't going to tell Jim what happened, he'd only have said 'told you so!' Though I still had no idea why he went on and on about birds and bees and puddings and buns!

****

FIFTEEN

Still smarting from my humiliating first date and vowing to give up romance for ever, I turned to my girls, especially Josephine, for solace. Helping Masie feed the calves was a joy and my heart mended almost immediately.

Happily letting the first cows in for milking the next afternoon, I was horrified when, one after another, they lifted their tails and let go great rivers of almost liquid greeny-yellow dung that cascaded over everything in sight, including me.

'Oh Jim,' I cried, 'they've got diarrhoea. Shall we call the vet?'

'Don't be daft, Fanny,' he said, laughing, 'they're scouring; happens when they get onto spring grass. They'll adjust.'

Milking became a nightmare trying to keep udders and teat-cups clean with endless buckets of fresh water and cloths, while enduring constant showers of obnoxious liquid effluent and dung-laden tails flicking across my face. I hoped it was good for my complexion.

Cleaning up afterwards took twice as long, and I couldn't blame Doris for charging me double for extra filthy dungarees and shirts. But my *very* unpopular girls did adjust and life got back to normal.

Soon after, while feeding the calves, I noticed a few very small scabs on some of their foreheads and noses and asked Masie what they were.

'Oh blimey, gal,' she said, looking closely. 'I think they've got the ringworm. We'll have to tell Jim. Don't touch them.'

Ringworm? I'd never heard of it. Was it like smallpox that people used to die of a long time ago before vaccination?

'Soon cure ringworm, Fanny,' Jim told me. 'got my own method – scrubbing with Quink ink – it's got to be Quink. Don't need those fancy remedies the vet prescribes. I'll show you how to do it.'

'*Quink ink?*' I was flabbergasted.

He came over to the calves that evening and while I held each affected calf as still as possible, he dabbed the scabs with ink and scrubbed them thoroughly with a nail brush.

'Do this every day and they'll soon disappear,' he told me. '*Don't scratch yourself and wash your hands thoroughly afterwards. Ringworm's very catching.*'

Masie and I found it difficult holding struggling calves while we took it in turn to scrub the scabs. Some grew bigger so we had to scrub harder, leaving sore patches – poor creatures. We made sure we thoroughly washed our hands afterwards, and Masie said she'd never caught ringworm.

Undressing a few nights later, I was shocked to see a very tiny scab had appeared under a bra strap. It didn't look any worse next morning so I ignored it, hoping it would just fade away.

I said nothing. Jim would be scathing about me catching the ringworm. Anyhow, it might disappear. But the scab began to multiply, thickening up, quickly spreading across my right chest. Now I was frightened, but somehow daren't tell anyone. Luckily I was having a day off soon; I'd go and see a doctor.

I might have guessed Mother's reaction. She shrieked in horror and backed away, bringing my three aunts at the trot – who never shrieked – but clamped their hands over their mouths and also backed away. I know I looked as though I'd got the dreaded plague, but I had hoped for a little sympathy.

Mother took me straight off to see the doctor. By now the ever-growing ringworms had almost taken over half my chest – curiously resembling a map of Britain, with Cornwall beginning to curl down round my right breast and the Scottish Isles marching up the back of my neck.

Even the doctor appeared to back off; quickly prescribing a large tube of something to rub in and ushering us out rather quickly. I felt I should ring a little bell and call out 'unclean, unclean'.

At home I was put into quarantine. I must keep to myself, stay away from the table and not come home again until I was cured.

Poor Mother, torn between her instinct to look after me, but anxious for me to go before I contaminated the germ-free, spotless house.

Still not telling Masie and Jim, I plastered my chest with the cream night and morning and turned my shirt collar up to hide my neck. But still the scabs multiplied, burning, thickening and itching – my right nipple now threatened by Lands End and John O'Groats dangerously near my hairline. Masie had told me if you got ringworm in your hair your head had to be shaved like a convict! I felt desperate. I'd have to confess.

'Oh, good Gawd, girl!' cried Masie, when she saw my modestly exposed shoulder. 'Jim, come here.'

'Good grief, Fanny!' he yelled, 'why the bloody 'ell didn't you tell us before, you nitwit. Let's see the rest of the damage.'

'*No*, only Masie,' I gulped, nearly in tears, and turning away I opened my shirt wide. I wasn't even able to wear a bra' now.

'Blimey, child,' she gasped, 'I've never seen anything like it!'

'Right Fanny,' Jim said, 'you'll have to scrub it twice a day with ink.'

Shaking my head, I told them I was already rubbing in the doctor's cream, though the scabs were still spreading. Anyhow, ink was for the calves, not humans.

'Forget that rubbishy cream, Quink ink's the only thing,' he insisted.

Masie found me an old nailbrush and with a bottle of ink I crept back to my room determined to get rid of the loathsome ringworm, imagining my whole body unrecognisable and having to be shut away in the Pest House! As it was, I kept my distance from Billy, Susan and Peggy and daren't go near Flora or her baby.

I had real empathy now with the calves – when I wasn't feeling sorry for myself – as gritting my teeth I tentatively scrubbed ink into my scabs. Results – nil, apart from eye-watering, jaw-clamping soreness, a blue-black chest and a deepening depression.

Secretly admitting my failure to Masie, she said she'd have a go, so after breakfast we closeted ourselves away in her bedroom and she tackled my chest with the same energy she used on her kitchen floor. But her confidence soon waned when I started to squirm and pull away, my watering eyes turning into rivers of pain.

'What we going to do, gal?' she said. 'I hate hurting you, but you've got to get rid of it. Jim'll be asking.'

He did.

'Hopeless pair of ninnies, you are!' he said. 'No use whingeing, Fanny. Got to grin and bear it. I'll do it.'

'Oh *no*, you can't,' I protested, my face flaming.

He laughed. 'What's the fuss? I've seen plenty of titties in my time.'

'Oh have you, Jim Larchford?' Maise said. 'What *don't* I know?'

'Before your time, Masie girl,' he said, grinning.

So, despite my horrified protests and with the awful prospect of me ending up as one giant scab isolated from life for ever, I consented.

With Masie standing behind me holding a very narrow scarf around my nipples to preserve my modesty, Jim went into vigorous action, dabbing and scrubbing, ignoring my squirms, squeals and pathetic tears – treating me exactly like one of the calves.

Inevitably, the scarf slipped from time to time. Scarlet faced, I'd shriek, Masie giggled and Jim shut his eyes in mock embarrassment. I *knew* he was enjoying himself. To protect my sore weeping chest, Masie gave me a large piece of lint to lay on it under my vest, and after two long, miserable weeks of torture-by-Jim, the scabs finally disappeared – but left very visible marks.

Terrified I'd be scarred for life, Jim reassured me they'd fade in time and I'd be immune from ever catching ringworm again.

Liberated at last! I felt ecstatic, almost light-headed – like my girls heading eagerly for the succulent Spring grass after the long enclosed winter. I'd be welcomed at home on my day off, able to meet up with my friends, and go to the cinema.

With the sun still shining, life was perfect – and all thanks to a bottle of Quink ink and Jim's savage administrations!

****

SIXTEEN

I was at last to find out what Jim meant when he told me Romulus 'did most of the business now, but Remus hardly ever', so when he said 'we're going to give old Uncle a treat', I thought it would be an extra long walk around the paddock and sincerely hoped it wasn't going to be *me* who had to take him.

Picking up a couple of halters from the harness room, (still the ghost of Paddy lingered), Jim told me to put one on Bella, the other on Petal and hold them back when Alf took the cows back to pasture after morning milking.

Impatient at my fumbling efforts to halter them and anxious to follow their sisters, Bella and Petal mooo-o'd indignantly and rolled their eyes at me resentfully.

'It's not my fault,' I told them, 'I don't know why either!'

It was a dreaded dunging-out day for Romulus and Remus, so I reluctantly presented myself in their paddock to find Jim already giving Romulus his daily constitutional.

'What the hell have you been doing, Fanny?' he shouted. 'Get cracking, before this fella falls down dead with exhaustion. I've been round the bloody paddock four times already.'

Muttering to myself *'you're a bloomin' liar, Jim Larchford!'* I mucked out and re-strawed Romulus' stall in double quick time, with just enough strength to do the other one while Remus had his airing.

But instead of putting him back in his stall, much to my surprise Jim led him towards the yard.

'Don't look so gormless, girl. I want to get a few more good calves from him before he's pensioned off. Young Romulus can go out in the field with his harem but Remus is too old and crotchety to be let loose. We'll start him off on Bella and Petal. Chop-chop, get lively!'

Well, I remember from history lessons that the Age of Enlightenment covered about one-hundred-and-thirty years – say from Galileo to Isaac Newton – but mine took less than *five minutes*!

Wide eyed, with flaming cheeks, my shocked brain took in the whole astonishing process of procreation, and more to the point, what Remus' treat was!

'You can close your mouth now, Fanny,' Jim said, laughing at my stunned face. 'About time you woke up to the facts of life.'

Next day Romulus went eagerly into the field with the cows and collecting them for milking became a two-man job, with either Alf or Jim coming with me armed with a stout stick to stop Romulus thinking he could follow his wives into the milking shed.

He chose an unlucky time to leave his warm, dry stall. The long sunny spell came abruptly to an end with sudden thunderstorms and heavy rain. Romulus and his ladies huddled miserably together near any shelter they could find and his top priority was certainly not treats!

Work didn't stop for heavy rain and Jim showed me how to use a rip-hook to rid a Lucerne crop of rogue thistles before it was cut for silage – whatever that was?

'I warn you, it's bloody sharp,' he said. 'You're no good to me with bits missing.'

Using the rip-hook was more dangerous than I imagined; trying to swing it so it cut the thistles but missed most of the Lucerne and any fingers or toes that might get in the way! I found half-bending jobs back-breaking, and as the cold rain soaked through my old school mack I remembered the excruciating aftermath of the turnip hoeing!

Jim told me silage was a high-protein winter feed for the cows – after storage in a compressed state – and we were making it for the first time.

One drizzly morning a lorry turned up and in the field next to the farm yard, a couple of men constructed a tall cylindrical object made of corrugated metal with ladders in and out each side.

Intriguing; but I had to wait to find out. Jim got impatient with too many questions.

The only person pleased with rain after the dry spell was Roy Pengelly, conscientiously tending his burgeoning vegetables and fruit bushes. Occasionally I went to inspect progress with beautiful Flora and her pretty little replica, Lily, who grew daily, rain or shine!

Flora's nature matched her looks and she seemed oblivious to continuing tight-lipped disapproval from Doris and Ethel. As for Colonel Fortescue's frequent offers of lifts into town – he hardly ever spoke to me now – she told me, with a laugh, she always refused.

In a bright spell at the end of May, Bert cut the Lucerne and a field of long grasses. Jim said both would be allowed to wilt for a few days to get rid of some of the moisture.

Dolly and Molly were reluctantly persuaded from their field and harnessed up to carts; then it was all hands to pitchforks. Great fun, tossing Lucerne up onto one cart and grass onto the other – a wonderfully aromatic contrast to dunging out bulls.

'Go and scrub your boots off, you three,' Jim said to me, Billy and Susan, who'd been helping too.

'Whatever for?' we asked.

'Just do as you're flippin' told,' he growled at us. 'You'll find out.'

I watched, fascinated, as the men alternately layered about a quarter of the Lucerne, then grasses into the cylindrical silo, patting it down firmly. From a huge tin, Jim poured a thick, black treacle-like substance – molasses – as evenly as possible over the top, then another layer of Lucerne and grasses.

'Now, young 'uns,' Jim said, 'hop up the ladder and tread all that lot down, hard as you can. Got to get rid of every bit of air.'

Suddenly I was a kid again, giggling my head off with Billy and Susan as we stamped round and round, feeling us sinking lower and lower as the Lucerne and grass compressed under our weight. We didn't mind at all having to repeat this three more times until the silo was completely full and it was then covered tightly with a tarpaulin.

On my next day off at home, when Mother and my aunts asked me what I'd been doing on the farm, I told them what fun it was making silage, but chose not to mention the delicate matter of my Enlightenment. I didn't think it a suitable subject for the Sunday tea table.

✳ ✳ ✳ ✳

SEVENTEEN

I'd grown accustomed to an ever-changing procession of my girls passing through the milking shed. Old friends dried out and pastured off for a couple of months before they calved again, to be replaced by nervous, fidgety first-calf heifers – new names, more bruises – then old timers returning. Milk yields fluctuated week by week, but overall I was told, averaged out during the year. The Ministry of Agriculture rationed food stuffs according to yield so the monthly figures were anxiously watched by Jim and the Colonel.

Happily on holiday with her other dried-out sisters, I visited Josephine most days for a cuddly chat, and in early June was thrilled to play assistant midwife at the birth of her third calf – but my joy was short lived. I'd imagined her producing a pretty little heifer I could spoil as she grew up, but alas the new arrival was a male calf and I knew what his fate was. Still, after a few days, Josephine was back in the dairy and we both enjoyed more togetherness.

Mrs Bartholomew called to see me and said as I was to be eighteen this month I would get a pay rise of two shillings and I'd be paid for overtime, plus a week's annual holiday at my employer's discretion. Perhaps I didn't look delighted enough because she emphasized what an improvement this was on early W.L.A. Terms of Employment.

I hastily thanked her and said I'd probably save the two shillings, as apart from going to the cinema occasionally and a bit of toiletry shopping, I had little chance to spend my wages, not even my sweet coupons! I'd forgotten what my working hours should be and couldn't imagine Jim giving me more than two days off occasionally, let alone a whole week.

'I hope you're going out to celebrate your birthday?' Mrs Bartholomew asked, seeming concerned I had no social life. When I shrugged, she told me if I couldn't have the whole day off she'd arrange an evening outing for me. I just hoped it wasn't tea at the local vicarage. I'd got rather fed up with vicars and curates coming to tea at home and tucking greedily into our rations; butter on *all* their bread and an extra large slice of cake.

With Romulus vacating his stall for the summer, Jim decided we'd give it a good clean. Why didn't I realise by 'we' he meant *me?* As he leisurely walked Remus round and round the paddock – keeping one eagle eye on me – I lugged out every last bit of eye-smarting, obnoxious dung from Romulus' stall, even scraping it off the walls, and then thoroughly hosed everywhere down.

'While I've got Remus out, you can carry on and do his stall, Fanny', he said cheerfully, ignoring my dismayed face. 'I'll stick him in Romulus' stall and go and mix up some whitewash. We'll slap a couple of coats on the walls; then when they've dried we'll bed up.'

With mutinous mutterings and taking the *'we'* with a pinch of salt, I struggled on, ignoring Remus' noisy disgust at being displaced, to just finish by the time Jim came back with the whitewash. He took pity on my weary, dishevelled state and did the whitewashing himself.

A spanking new tractor was delivered – a shiny blue Fordson Major with giant wheels; twice the size of the shabby little red Ford. Bert's usual grin was twice as wide as he pointed out all the modern improvements and told us how much easier it would be to drive.

Perhaps now, I wondered, Jim will let me have another go?

'Don't even think it, Fanny,' he said, reading my mind. 'Broken bridges and gateposts are too costly.'

A wild exaggeration! I hadn't even hit *one* gatepost. But maybe, seeing my disappointment, he might give me some time off on my Birthday, the Eighteenth?

'Not a hope, girl,' he said. 'Soon as this rainy spell's cleared we've got to get the hay in. Any day now.'

Mrs Bartholomew called soon after to tell me she'd arranged for two land-girls who worked on a Winchfield farm to take me out for the evening. So I accepted, but immediately worried I'd no decent dress to wear. My hands were permanently grimed, my nails broken, my face too ruddy and my hair dull. Ah well, this was the price I paid for Serving my Country. I only hoped the other girls looked the same.

It was as we finished morning milking on the 17th I suddenly realised Josephine hadn't been through. Alerting Alf, who simply shrugged and carried on scrubbing floors, I searched the yard, then, very worried, ran all the way back to Upper-gate pasture, where the milkers were grazing.

No sign of Josephine, but staring at me balefully was Romulus, daring me to venture into his territory. I climbed two bars of the gate and from this better view-point I spotted Josephine, laying flat beneath the far hedge.

'Oh *no*!' I moaned. Was she dead? I must be brave; go to her. My heart thumping, I climbed over the gate and slowly walked towards her. At first Romulus didn't move, so I kept walking. He started grazing, and plucking up courage I quickened my pace, softly calling 'Josephine, Josephine.' She lifted her head. What a relief, but she must be ill or injured.

Unfortunately Romulus lifted his head too and deciding I was taking unpardonable liberties, began trotting towards me, bellowing.

To my unforgivable-unforgettable shame, my love for Josephine was instantly overridden by my intense fear of Romulus and I bolted for the gate, just throwing myself over before he came snorting up.

I rushed to fetch Jim, and Romulus was soon held back by Bert while we went to examine Josephine.

'Umm, looks like she's injured a back leg – or broken it,' Jim said, after he'd failed to get her to stand.

I was almost happy when Mr Blunt – who reminded me too much of his heavy-handed doctor brother – pronounced a bad sprain, not a broken leg, which would have meant her being put down. Grunting with pain, she was carefully lifted onto a cart and trundled by Bert's new tractor to a freshly bedded-up stall to rest. Jim stripped her milk off as her udder was painfully full, but he said she'd dry off fairly quickly now. I was fiercely adamant I'd take over her care, though Jim said he and Alf would have to lift her daily to check the leg. I could change bedding, and feed and water her.

In my concern for poor Josephine, I'd completely forgotten my birthday outing. I didn't want to go, but Masie and Jim insisted I should, persuading me to stop worrying about Josephine.

'Go and kick up yer heels, Fanny,' Jim said. 'But no hanky-panky, mind!'

Nervous but excited, hoping we might go to the cinema, I was called for by the two girls, luckily both in uniform like me, who introduced themselves as Joan and Monica. They'd arrived in an old van and as Joan hopped into the driving seat, Monica opened the back doors and said 'jump in, the boys'll squeeze up'. To my horror I was grabbed by four massive hands and hauled inside, to find myself wedged between two grinning soldiers!

'Hello Jeanne,' one said, 'I'm Tony, this is Phil. Happy Birthday!'

Scarlet faced, I was totally tongue-tied. Mrs Bartholomew never said there would be *soldiers!*

'Where shall we go tonight, fellas?' Monica shouted over her shoulder.

'Let the Birthday girl choose,' said Phil.

'Oh, er – the cinema?' I whispered hopefully, (I wouldn't have to talk much in the dark).

'Ah, there's nothing worth seeing at the flicks this week,' Tony said. 'Let's give The Kings Arms a go and the others cheered.

I suspected The Kings Arms was a *Public House!* I'd never been in one. 'Not a suitable place for nice young ladies' I heard my three aunts chanting – but what could I do? I had no choice.

Trying hard to pretend I was a seasoned pub crawler, I followed the four into the bar and looked about with trepidation. Apart from a couple of harmless-looking farm workers, it was empty. The landlord had a jovial sort of face. It didn't look like a den of iniquity so I relaxed a bit, until Tony said 'What'll you all have? First orders on me.'

While the men were ordering beers and the girls *cocktails*, I frantically scanned the bottles for something safe. I'd never drunk alcohol, but I'd look babyish ordering lemonade. I spotted a big barrel on the bar counter with pictures of apples on it labelled 'Cider'. Sounding confident, I said I'd have half a pint – after all, apples are quite harmless.

As I sipped my refreshing glass of bubbly apple juice, I envied Joan's and Monica's easy conversation, but gradually found myself not feeling quite so shy; suddenly able to join in and even laugh at jokes – which I mostly didn't understand, while they laughed at everything I said, though I didn't think I was being funny? When Phil said 'second orders', I asked for another cider. The evening was going splendidly; I couldn't stop smiling. Who needed alcohol?

Eventually the landlord yelled 'last orders' and Tony said 'Whoops, we'd better get going. Got to be back in camp by ten thirty.'

I stood up to leave – but couldn't! My lower extremities were missing. I checked; my legs were still there. How strange! I tried to stand again. No luck. The room began floating around me; people seemed to have two faces.

'Crikey! She legless.' said Phil.

'What's wrong with me?' I moaned, desperately trying to get up.

'You're just a wee bit tiddley, darling,' Tony told me and they all roared with laughter – but it wasn't funny.

'Come on, we'll help you,' Phil said, and he and Tony grabbed me under the arms and part carried, part dragged me outside and into the back of the van, where I sprawled, utterly confused. How could I be drunk – *a mortal sin* – despite only drinking apple juice?

The journey back to the farm passed in a swirling blur, until Monica and Joan opened the van doors and I tried to climb out, but still couldn't, much to everyone's amusement.

'Don't fret, darling, we'll carry you to your front door,' the boys offered.

Mortified, I let the still giggling girls pull me out of the van and lift me onto the boys' crossed arms, and we processed, much too noisily, up the drive.

'Ssssh!' I begged. It was getting late and Mrs.Fortescue would be in bed. Unceremoniously, still stifling laughter, the boys dumped me in the front porch. I clung desperately to a pillar.

'You'll be tickety-boo in the morning,' they reassured me. 'Don't fret. 'Bye!'

'What a joke!' I overheard Tony say as they disappeared down the drive, still laughing.

It wasn't a joke, crouched in the porch, wondering what to do. I tried the front latch. It was fastened shut. I knocked lightly hoping Colonel Fortescue was still up. He didn't answer. I knocked again, louder. No luck. I'd have to somehow get round to the scullery door, hoping it was still open.

With tingling legs just beginning to wake up, I half crawled, half swayed around to the rear of the cottage, but that door was locked too. I knocked, loudly this time. Still no answer. I began to panic. Then I realised the Colonel might still be in his library, the window just round the corner. To my relief I saw him in his armchair, an empty glass at his side, his eyes closed. I'd have to wake him; he wouldn't be pleased.

I tapped on the window lightly. He didn't move. I tapped again, hard this time. Still he didn't wake. I tried once more. He sat like a dead man.

Now I really did panic; it was rapidly getting dark, beginning to drizzle. I was shivering. I didn't dare wake Masie and Jim – he'd give me hell! I didn't dare wake anyone – until a sudden vision of a warm, straw-bedded stall gave me the answer. *Josephine*. She'd be pleased to see me.

With legs that now worked but still had their own agenda, I lurched across the field, straining to see the farm path in ever-increasing darkness. I managed to grope my way to the sick-bay shed, but it was bolted. I should have known. I could try a window – the one behind Josephine's stall was slightly loose. After a nail-breaking struggle I managed to prize it open and heave myself over the sill, landing in an ungainly heap onto the straw below – but luckily not on Josephine.

I heard her grunt before I gradually picked out her shape in the near darkness.

'Josephine,' I whispered, 'don't be afraid, it's only me.'

She gave a tiny anxious 'moo', so to reassure her, I stroked her head and back. Stretched out as she was, I had only a small irregular area I could settle in. Damp and shivery, wishing I had my greatcoat, I somehow persuaded my body into an uncomfortable triangular shape, Josephine making up the third side – Isosceles style – and pulled as much straw as I could over me.

I waited for blissful sleep, but none came. I hadn't bargained on Josephine's noisy digestive enzymes working an energetic night-shift or her frequent explosive contributions to global warming, interspersed with grunts, gurgles and much tail swishing. Nor had I realised there would be tiny, unwelcome 'others' seeking a dry straw bed on this wet, cold night.

Morning found me stiff and exhausted, covered hedgehog-like with bits of straw, and no time to change or tidy up before fetching the cows.

Alf just stared, but when Jim came into the milking shed, I braced myself.

'Stone the crows, Fanny! Didn't I tell you – no hanky panky, no rollin' in the straw. Blimey girl, what'll I tell your Mum if you've got yourself in the puddin' club, eh?'

Scarlet-faced, indignant, I tried to explain, but he didn't listen.

'And another thing,' he lectured, 'she wouldn't like you going in pubs and getting drunk.'

If I could have got a word in, I'd have told him I had no intention of ever going in a pub again, or being taken out by land-girls with *soldiers* – and I certainly wouldn't dream of visiting one of those pudding clubs!

****

EIGHTEEN

Jim promised me a day off after we got the hay in, so I willed it to stop raining, but where was flaming June?

In the meantime I concentrated on Josephine. Despite rest and some gentle leg massage recommended by Mr Blunt, Josephine still couldn't stand, which puzzled him – and worried me. She'd nearly dried out now and without exercise wasn't hungry, even though I tempted her. She'd look at me with gentle, puzzled eyes that nearly broke my heart.

What a relief when one morning I woke to bright sunshine. Jim had been short-tempered, so when the wireless informed us we were in for a dry spell, he cheered up, and so did I.

As with silage making, it was all hands to haymaking. Proudly driving the new tractor, Bert cut the two large fields of grass, which was left to dry out a day or two before it could be stacked.

At last I felt like that Happy Healthy Land-girl of those W.L.A. Posters! With Masie and Flora, (in eye-catching red shorts), we lifted and turned the hay with giant rakes ready for picking up. Warm sunshine on our backs, birds singing and lots of laughs made heavy work fun. Lily watched contentedly from her pram and the Colonel checked up on the red shorts from time to time.

With Molly and Dolly harnessed into carts, everybody toiled through each day until dark, tossing great pitch-forks of hay up onto the carts. Susan and I were given the enjoyable job of riding aloft to level the loads, while Billie led the horses – destination, a rectangular area near the farm buildings where Bert and Jim would build the haystack. Hungry and dry mouthed we downed great mugs-full of thirst-quenching cold tea and appeased our growling stomachs with giant cheese and pickle sandwiches.

What those deceptive W.L.A. posters didn't show was the bone-aching weariness at day's end. Ghost-like from a thick coating of dust, I'd peel out of my clothes and plunge my sore, itching body into a hot bath, to nod off until cold water woke me, shivering, and I'd crawl into bed to a blissful, dreamless sleep.

Josephine still wouldn't stand, so Jim called Mr Blunt. Bringing a harness contraption, they hoisted her up, but still her leg gave way and she grunted with pain. Small pressure sores were developing where she'd lain so long. Mr. Blunt prescribed some cream to apply each day.

'She's a puzzle,' he told Jim, 'but we'll give her another week; then – '

Then what? It sounded ominous. Some vet! This Mr Blunt was as useless as his doctor brother was unfeeling. Why couldn't he cure Josephine? Now she was getting worse and I felt so helpless.

Jim showed me how to clean and treat the sores, squeezing any puss out before applying the cream, but Josephine kept tossing her head with pain, looking at me piteously however gentle I was and despite my explaining how much I hated hurting her.

At last my promised day off arrived, though I wouldn't have left Josephine if Masie hadn't said she'd care for her.

I'd saved enough clothing coupons to buy a new dress, but had to use the rest up on a second pair of dungarees. A new, shiny-looking material called 'plastic' was in the shops – *off* coupons, sold from the roll, and my friends were choosing the bright pastel colours and making triangular headscarves. I bought half a yard of green, enough for two scarves.

My New York sister sent Mother some women's magazines full of smart designer clothes with unusual fashion touches and *no* skimping of material, so unlike my new mauve-striped dress with its small, plain sleeves and flat, uninteresting skirt.

My pleasure of a new dress was soon snuffed out; Jim had called the vet again. The week was up and Josephine no better. I knew what they would decide and I couldn't bear it.

'Stop blubbing, Fanny,' Jim said, not unkindly. 'It don't do, girl, to get so attached to a farm animal. Cows aren't pets; I did warn you.'

Horrid Mr Blunt arrived and I was ordered off to help Masie with the calves. It was the cruellest day of my life on Brinkleigh Farm.

Next evening, to cheer me up, Jim said he'd take us all on a picnic the first fine weekend to a nearby lake where we could swim. It sounded fun, but what could I swim in? I only possessed a shapeless, moth-eaten school costume I wouldn't want to be seen dead in. I wished I hadn't used up all my coupons. Then I remembered the no-coupons plastic material I'd made into scarves. A page full of glamorous American bathing costumes flashed into my mind and I knew what to do! I'd copy one of the designs and make it myself. I remembered one costume distinctly and could cut my own newspaper pattern.

Begging a lift into Fleet with Jim one afternoon, I headed for the drapers shop. Luckily, they were stocking two colours of the plastic material – bright pink and pale blue. Choosing the pink, I was assured it could be machine stitched. Knowing Flora had a sewing machine, I timidly asked if I could use it. Agreeing, she seemed very keen to see my design.

'A *two* piece!' she gasped, 'in this Country?' She laughed. 'Why not? Go for it, girl!'

Realising it was very daring and modern, I spent every spare minute cutting out, modifying and stitching this unique creation. I needn't have hurried because summer had completely disappeared, replaced by rainy, stormy weather that didn't invite picnics.

'You look just like a film star!' Flora cried, as I paraded in front of her in my designer two-piece swimsuit. Admiring myself in the mirror I had to agree!

The bra top was tied round my neck with a big bow, and another bow, equally large, fastened it behind. Then appeared that shocking gap – *bare flesh* – down to waist-level pants, also fastened on each side by three more sizable bows. I was an amazing bright-pink confection! But did I really have the nerve to wear it?

No time for second thoughts because the sun emerged again and Jim and Masie fixed next Sunday for our picnic.

Rushing through afternoon milking as quickly as possible, we all squeezed into the old farm van, me with my secret bombshell wrapped in a towel – and *serious* second thoughts!

Jim drove us to a large wooded lake a few miles away and we settled down on a pleasant grassy bank. I noticed with alarm there was a group of soldiers bathing further up the lake. My serious second thoughts leapt into first place.

Billie and Susan, always hungry, wanted to eat first, but Masie was adamant 'swimming after food is dangerous', so they rapidly changed into their bathing costumes and urged me to hurry up.

Disappointed Jim and Masie weren't going to swim, I reluctantly found the nearest bush and hands trembling, tied on my oh-so-revealing swimsuit. Wrapping the towel round me I joined Billie and Susan, already dipping their toes in the water.

'Get in there, Fanny – you won't drown,' Jim called. 'I want my tea.'

I had no option. Gingerly, I dropped the towel and face burning, stepped into the lake.

'Good God!' gasped Masie. 'you're half *naked!*'

'Blimey, Fanny, you're making my eyes water,' cried Jim. 'You look like a bloody birthday cake.'

'Get in the water quickly, girl,' urged Masie, 'those soldiers will see you.'

With Billie and Susan staring at me open-mouthed, I quickly plunged into the breath-catching cold water and striking out purposefully, (but keeping one foot on the bottom), I pretended to be the expert swimmer I most certainly wasn't. I heard a lot of wolf whistles from the soldiers and hoped they'd go away before I came out again.

Bobbing around with Billie and Susan, I started to warm up and enjoy myself – until a strange, heavy sensation began to tug me down. Horrors! Water was gushing *into* my pants and bra top. Frantically I tried to press it out, but I was expanding rapidly – like a giant balloon! Suddenly, my pants POPPED, then the bra cups – ONE, TWO! My beautiful creation was disintegrating. My agonised squirms only made it worse. Pink plastic shreds and ragged bows began bobbing up to the surface all around me.

Hardly a snippet left to cover my bare essentials! I let out an anguished howl. Was there a nearby hole I could vanish into?

Masie rushed to the water's edge.

'What's wrong, girl,' she cried. 'Have you hurt yourself?'

'Her bathing costume's all torn off,' Susan said, holding up a break-away bow.

'It looked jolly daft, anyhow,' scoffed Billie.

'Come on Fanny', yelled Jim, 'stop playing silly buggers. Just get out. I'm hungry.'

'I'll hold the towel at the ready,' said Masie, paddling into the lake.

'I can't,' I wailed, 'I'm bare!' Not *strictly* true – I did have two bows still holding up my ragged waistband and a third draped round my neck with a bit of a bra dangling down.

'Turn your backs, you lot,' ordered Masie. 'Now, make a dash for it, girl.'

I couldn't; *wouldn't!* The soldiers were still there taking a great interest in me. Masie was getting impatient, Jim furious and the children hungry and fed up. But I just could not summon up enough courage and despite the cold creeping up and numbing me, I crouched submerged up to my neck.

Suddenly Jim's voice bellowed out.

'IF YOU'RE NOT OUT BY THE TIME I COUNT FIVE I'M COMING IN TO DRAG YOU OUT MYSELF. ONE ... TWO ....'

A threat I dare not ignore! I rose up out of the water like Botticelli's Venus – but a lot faster and minus the shell, and flung my glistening nakedness into Masie's outstretched towel.

I heard cheering and wolf whistles from the soldiers, just as Jim got to '... FIVE' – but he'd already turned round – the rotter!

It was after the hungry Larchfords had eaten enough tea that the sniggers, the giggles, the laughter took them over and I had to sit there covered in acute embarrassment, wishing this was a nightmare and I'd wake up soon.

'Don't look so glum, Fanny,' laughed Jim, wiping his eyes, 'see the funny side.'

How could I? My humiliation was bottomless.

****

NINETEEN

The only person who didn't snigger over my mortifying experience, but was really sympathetic, was Flora. She'd just read that cotton tape should be used to reinforce plastic seams. Why wasn't I told? Too late now. I never wanted to hear the word plastic again and hoped Jim would soon stop teasing me and shut up about it.

I needn't have worried. Something suddenly happened to put everything else out of the whole British population's minds.

BANANAS! In the shops again, in limited supplies, and everyone queuing up for their first taste after seven long years. Billie could just remember them, but Susan didn't and he hoped she wouldn't like it so he could eat hers.

Then something happened at Brinkleigh Farm to take all our minds completely off bananas.

Roy Pengelly's flourishing garden – his pride and joy – was broken into; the culprits discovered unashamedly still there, rapidly devouring every single green leaf and shoot and what they didn't eat they stamped all over. Who else could it be but greedy Big-un and his mates!

I was glad I couldn't be accused of leaving the sty door open. Everyone looked at Alf, but even he wasn't to blame. Big un had broken the sty door down and with his fellow conspirators in close pursuit, had headed straight for a very early Harvest Festival luncheon.

Roy was inconsolable, the Colonel livid, and metaphorically putting on his black cap, he passed sentence on ringleader 'Big-un, to be taken to a place of execution and instantly transformed into joints, chops and sausages' – half for the Ministry of Food and half for the Colonel. Despite my usual soft heart, I didn't feel one twinge of pity.

Flora, Masie and I found time to help Roy tidy up the ravaged garden but so much had been lost and although he replanted some vegetables, it would depend on a long hot summer for good results.

What summer? We hadn't really had one yet, only rain and more rain. Jim listened avidly to the wireless weather reports, and farming programme predictions for this year's harvest were extremely gloomy. As a precaution, the Ministry of Food had now rationed bread and flour, though manual workers would get extra. Masie, whose meals satisfied us fairly well, was moaning that she didn't now know how she'd manage to fill our hungry stomachs.

Despite the weather, Bert had ploughed, and was now harrowing, another big patch of the same field and Jim totally mystified me by saying he and I would 'broadcast' the seeds by hand. Meanwhile, I was to help him prepare a couple of cows for a coming Agricultural Show the Colonel was entering them into.

While Jim was choosing two of Remus' finest offspring, first-calf heifer Ella and a three-year-old, Dorcus, he suddenly swore loudly. Not unusual, but puzzling.

'Sod it, Fanny, we've got warble-fly! Luckily not on these two. Bit late in the season though.'

I couldn't see any flies, certainly none warbling.

'Don't look so dim, girl,' he said. 'they aren't buzzin' about – yet. Come over here, chop-chop. See these bumps on old Duchess' back – the warble fly deposits her eggs on the legs, they hatch out, burrow into the body and imbed themselves in the cow's back under the skin.'

'Horrible!' I shuddered. 'How painful!'

'Yeah, you can say that. The growing maggots cause the swelling, bloody things. Besides, they ruin the animal's skin if they mature and pop out – it's worthless afterwards'

'What can you do?' I asked him.

'Get rid of them, pretty damn quick. I'll show you how to squeeze 'em out after we've first checked every cow'.

Luckily only three, Duchess, Bella and Ruby were infected and after the next milking, Jim haltered them in the yard and showed me on Duchess how, with thumbnails pressing down really hard into each side of the lump, a fat, squishy maggot, moist with yellowy pus, slowly oozed up through the skin. It made me want to vomit and not surprisingly, Duchess grunt loudly and stamp her hooves.

'Must squeeze out *all* the pus; got it, Fanny?' said Jim, throwing the maggot into a bucket. 'Three more on Duchess, then tackle the others. Keep you busy 'til supper, eh.' With a grin he was gone.

I knew better than to call out 'can't do it' and was much too gentle with patient old Duchess, realising I caused her more discomfort than if I'd gone in with 'both thumbnails blazing'! Bella didn't take kindly to my hesitant surgery, landing me a kick or two for her efforts. By the time Jim returned I hadn't tackled Ruby yet, felt very queasy and fully expected the usual tirade about my shortcomings as a land-girl. But none came.

'Well done, Fanny,' he said, looking surprised. 'Didn't think you'd be up to it. '

Did he really mean he hoped I *wouldn't*? But I hugged that bit of unexpected praise to me and felt ten inches taller.

Ella and Dorcus were introduced to being led around on halters and standing quietly when told. Dorcus was a quick learner but Ella decided it was a gross infringement of her bovine rights and took much more persuading and a little bribery, before she finally condescended to humour us and emulate her sister.

Between milkings the two were now housed in the hospital-wing-cum-Pampering- Parlour and I helped Jim and Alf give them the full beauty treatment. Head, body and tail shampoos, hoof scrubs, pedicures and polishing, currie-combing tangled tails – with an artistic snip here and there – followed by a long, soft brushing from head to foot until their coats shone like satin.

'They're bound to win' I said, admiring our handiwork and thinking how much more glamorous they looked than me, with my stringy hair, red hands and dung-splattered dungarees. 'What will the judges be looking for?'

'Ah, a pretty, feminine face, of course,' Jim told me, 'a level back with fine shoulders, strong legs and stance and oh yes, a good-shaped udder with well-attached teats.'

Positive that our two beauties had all these good points, I asked Jim if I could go to the Show to help.

'Not a hope, Fanny. Masie and I go with the Colonel and Mrs Fortescue. You and Alf can hold the fort.'

I guessed it would be a very silent fort with no Josephine or Paddy to talk to.

But it was worth it. The Fortescues, Jim and Masie came back triumphant. Ella had picked up a Third rosette in her class, and Dorcus a *First* in hers!

'Means a lot to Jim,' Masie told me, 'to know the Colonel's so satisfied with his work with the herd.'

I nearly said it means a lot to me too when Jim's happy. I hoped his good mood would last, and sure enough Alf and I got an unexpected appreciation for our grooming efforts. Very pleased with myself, I grew another two inches.

'Let's get those seeds in while it's not raining, Fanny,' said Jim next day, and at last I was initiated into this mysterious 'broadcasting' that was nothing to do with the wireless.

Instead of using a seed drill and tractor, Jim and I donned large aprons with roomy pockets in front – me feeling like a kangaroo without my Joey peeping out – but instead half filled with a heap of seeds, ground peas.

'This is the way seeds were always sown in the old days,' Jim told me, as we stood at one end of Bert's newly harrowed ground. 'It'll be a late crop – if the flippin' sun ever comes out this year.'

I looked up at the watery August sky. Summer was so elusive.

'Now, Fanny,' Jim continued, 'pin back yer lug'oles and copy me. Move forward when I do, always keep level, you on my right about five feet away.'

With his left hand he scooped up a handful of seeds from his apron, and flinging his arm out to the side, he gradually released the seeds into the damp earth. He stepped forward a pace, looked back at me and said, 'Come on, Fanny, catch up. That's the ticket. Loosen yer wrist a bit more, use your thumb and scoop out more seeds next time.'

Feeling awkward at first, I quickly fell into Jim's rhythm as side by side we moved systematically up and down the field, me having to take big strides to keep up with his long legs.

'Crikey, gal,' he said, grinning at me. 'You're a natural. Not as bloomin' useless as I thought.'

Could this really be happening? Jim and I working together in perfect harmony; peacefully broadcasting seeds as our ancestors had done for hundreds and hundreds of years. It was unbelievable.

I suddenly felt the happiest I'd ever been since I first came to Brinkleigh Farm.

Ah, but little did I realise, the sword of Damocles was hanging precariously over my unsuspecting head!

****

TWENTY

Given another day off, I went home in high spirits and enthusiastically regaled my Mother and aunts with the fascinating techniques employed in grooming cows, broadcasting seeds and annihilating warble flies, though I left out some of the details. I could see it was putting them off their tapioca pudding.

The girlfriends I met in the evening looked at me strangely when I tried to impress them with my riveting life. It seemed they were now only interested in film stars, lipsticks and which boys they fancied. I felt the odd one out.

Catching a very early train back to Brinkleigh Farm next morning, I arrived in time for morning milking with silent Alf.

I was just clearing up in the dairy afterwards while Alf took the cows back to Upper-gate pasture, when Mrs Fortescue, smart in a nautical outfit, appeared at the door.

Surprised, I said, 'Good morning' and looked at her blankly.

'I've come for the milk', she said, somewhat impatiently.

I went hot all over. Oh horror! She'd asked me to save her eight pints of milk for this morning. They were going for a weekend with friends on their yacht on the Hamble. It had gone right out of my mind – but then, she did ask me just as I was climbing on my bicycle to catch the train home two evenings ago. Why hadn't she asked Jim or Alf?

'Don't tell me you've forgotten?' she bristled, seeing my scarlet face.

'Oh, I'm so, so sorry, really I am,' I said miserably, knowing that I'd heard the milk lorry leave just five minutes ago. 'I've been on leave and didn't have time to write it down.'

'Don't make excuses.' Now she was really angry. 'You've *ruined* our weekend. We'll have to spend precious time trying to beg milk from somewhere else. *Extremely* careless! The Colonel will be furious!'

I began to repeat how sorry I was, but she marched off in high dudgeon, leaving me feeling awful and knowing Jim would soon hear about it. I'd be in big trouble.

I was. Despite my excuses and abject apologies, I was 'thoughtless ... lacking in concentration ... needing to pull my socks up ...' He went on and on.

It was unfair. Not true. Gone was my happy, new-found sense of achievement, of belonging, of working harmoniously with Jim. All because I'd forgotten the dratted milk.

It was a miserable weekend, no one to commiserate with. Flora was away with little Lily visiting her mother. Even Masie wasn't understanding.

'Everything that goes wrong on this farm, gal, comes back on Jim, see,' she told me pointedly.

Nosy Ethel, who cleaned for the Fortescues, delighted in driving the knife in further. 'She weren't 'arf mad with you when she came back from the dairy; the Colonel too.'

So when I saw their car return early Monday afternoon as Jim and I were walking up to fetch the cows for milking, I vowed to keep my distance.

We had almost reached the gate to the pasture when the Colonel shouted for Jim to go over and see him.

'I'd better go, Fanny,' he said, thrusting the stick into my hand. 'You can bring them in. Don't look so petrified, you've seen how Alf and I do it a million times.' With a wave he was off, leaving me stunned, dry-mouthed and sweating with fear.

Taking a deep breath, gripping the stick, I approached the gate where most of my eager girls were gathered, with Alice and Princess at the front as usual. Romulus was further back, paying attention to one of his harem, but as soon as I opened the gate and the girls began to stream through he ran to catch up.

'Oh, please God, *please*, make me brave,' I prayed aloud. 'I *promise* I'll be good for ever; *never* tell lies; go to church *every* Sunday if you'll just let me be brave today. Oh, PLEASE GOD, PLEASE!'

Resolved, I stepped forward brandishing the stick and as Romulus bounded towards the gate I managed a rather quaky version of Jim's voice, 'BACK, Romulus, BACK boy. OH, *FOR GOD'S SAKE, GO BACK ROMULUS!*'

But Romulus knew I wasn't Jim; I was just that chit of a girl he liked to terrify, so he came straight at me, head down, a mean gleam in his black, beady eyes.

With a last feeble wail, '*go back*, Romulus, *go back*,' I flattened myself against the gate as he swept triumphantly through with the rest of the girls. I'd failed, miserably. Now I *had* to keep him out of the milking shed.

Jim still wasn't back by the time the yard was full of hungry cows, plus Romulus, and I managed to let the first six in without a hitch. I got them on the machines and enjoying their tea. Romulus was exploring his new territory, so I was able to let a few more girls in and out, until he smelled their tasty cow-cake tea and pushed his way through to the door, bellowing loudly to be let in.

I was now in full panic mode. Shaking, I managed to finish off and let out the six girls already milked. Armed with the stick, I now faced my Armageddon. But Romulus thrust past me, nearly bowling me over, and squeezed into the first pen, loudly demanding his tea, while his ladies fidgeted in theirs, uncertain of his presence in their all-female sanctuary.

It was the ultimate disaster. I felt numb. In robot fashion, I got the girls onto their teat-cups and fed them, which made Romulus bellow even louder, so in desperation I gave him a hob-full to shut him up, deciding I'd push him through as soon as he'd finished eating.

Jim still hadn't returned. I prayed I'd get Romulus out before he did. At least my telling off would be one degree less – perhaps?

I got another two girls through and then, opening Romulus' pen exit gate, I gave him a big whack on his rump.

'Out you go, you horrible animal,' I ordered him. '*OUT, OUT!*'

Romulus' response was to crash his head against the pen bars and bellow loudly.

I shouted at him again and whacked him twice, harder this time – the same response.

His ladies didn't like me shouting and began fidgeting and mooing. A first-calf heifer, Heidi, kicked off her teat-cups and began backing out; followed by highly-strung Princess. Frantic, I rushed to persuade them back in, re-wash the teat-cups and get them milking before I tackled Romulus again.

This time I not only shouted, I prodded him with the stick, whacked him soundly and pushed with all my strength. But he had no intention of moving. Four hooves stubbornly planted, Romulus was exercising his squatters rights and announcing it loudly.

Now the girls were really upset, Heidi backing out again, teat-cups in the gutter, with Princess following, then Rosa.

This was the moment Jim chose to appear at the door.

'WHAT THE BLOODY HELL -?' he yelled.

'Oh Jim, I couldn't stop him; I tried, really I did.'

'GOOD GRIEF! CAN'T I LEAVE YOU FOR ONE MINUTE WITHOUT SOME MAJOR DISASTER HAPPENING?' he shouted. 'DON'T YOU KNOW THE DIFFERENCE YET BETWEEN A RUDDY UDDER AND A BLOODY BIG PAIR OF BALLS, YOU WITLESS IDIOT?'

'It wasn't my fault. He takes no notice of me,' I wailed.

'Hoping to milk him, were you?' he ranted, whacking Romulus on his rump. 'Don't stand there gawping then girl; see to those cows, God damn it, before they wreck the place.'

I jerked into frenzied action, all fumbling confusion as I tried to pacify the girls and get them milking again. Two were finished and I let them out. Should I let more in, or wait until Jim had got rid of Romulus?

But all Jim's efforts at whacking and pushing were coming to nothing; Romulus was a very angry fixture and with each fruitless effort, Jim's anger was reaching boiling point too. Unrepeatable expletives rained down on my head like volcanic ash.

'GAWD HELP ME, GIRL,' he roared. 'Go and get Alf, DAMN YOU, and Bert if you can find him. *RUN!*'

I found Alf quickly, but Bert was nowhere to be seen.

'You, Fanny!' Jim shouted. 'Get back to the bloody milking – and let those two beggars out, they're upsetting the others.'

With one eye anxiously on Jim and Alf as they pitted their combined weight against Romulus' unyielding posterior, I tried to create some sort of order out of chaos, but the girls were jumpy, nervous, holding back their milk, kicking teat-cups off. I was almost at my wits' end.

It couldn't have been a worse moment for Colonel Fortescue to poke his head round the door.

'Heard a commotion, Larchford – what's up? he asked, taking everything in. 'Good Lord, how did *he* get in here?'

'You may well ask, Colonel,' panted Jim, glaring at me. 'He's thoroughly stuck – or chooses to be. Can't shift the bugger.'

'I'll give you a hand,' the Colonel replied, and the three of them whacked, prodded and shouted, Romulus snorting and stamping but not moving one hoof forward.

I tried to pacify my milling, mooing girls as they pranced nervously in and out of their stalls, teat-cups and buckets flying.

Suddenly there was a cheer as Romulus decided he'd co-operate at last and squeezed himself slowly forward into the passage way to the outside. He was a tight fit and it took a lot more shoving and shouting before he finally shot out into the yard beyond, bellowing his bad temper.

In the immediate quiet, keeping my head down, I busied myself with the rest of the milking, gradually restoring peace and order. But I knew my moment of reckoning was fast approaching, making my stomach churn.

When Jim came back, grim faced, the expected explosion didn't happen! Silent, he saw to the churns, dismantled the cooler and inspected the Min. Ag. sheets.

'Huh, as I thought,' he spoke at last. 'The yields are well down. Not surprised; are you Fanny?'

Miserably, I shook my head.

'I'm taking the cows back,' he snapped, 'You get on and clean up – look sharp! Straight after, the Colonel wants to see you.'

I guessed what that meant – instant dismissal.

One hour later, clean and tidied up, I stood before Colonel Fortescue, feeling like Marie Antoinette about to keep her appointment with Madame Guillotine.

My heart was thumping so noisily I couldn't take in exactly what he said about my misdemeanours and shortcomings, but with an audience of two friends sitting nearby, I could tell he was enjoying himself as he thoroughly wiped the proverbial floor with me – the wronged employer, the useless land-girl. (I guessed he'd never forgiven me for not letting him count my ribs that evening and this was *pay-back time!*) I didn't try to excuse or justify myself, but crept up to my room, crushed, shamed, only surprised that he hadn't actually sacked me!

As I lay on my bed, rivers of tears wetting my pillow, I knew I just couldn't face anyone again. I'd been humiliated, often bullied, rarely appreciated. But misery was slowly overtaken by hot anger. So, I hadn't been sacked, but I didn't *have* to stay here. I wouldn't. I'd leave at once!

Commonsense told me I couldn't just walk out in broad daylight. So I made a plan. I'd pack up quietly now, wait until after dark – just after ten – sneak my bicycle out of the shed, load it up and walk to Fleet Station to catch the milk train home to Basingstoke. I'd have to put up with my growing hunger. I was quite excited!

Trying to fit all my belongings into the small suitcase was impossible. So I pulled on breeches over dungarees, one shirt over another, sweater over cardigan, socks over socks, overcoat, mack, hat and gumboots over the lot until I resembled Beano's Billie Bunter and could hardly stagger under the weight. I squeezed as much as I could into the case, found a canvas shopping bag and crammed it full to overflowing, tied my shoes together with their laces and decided I'd dangle those over my handlebars.

Then came the long wait, hour after hour, my stomach churning in protest, until it grew dark enough and I heard Mrs Fortescue go to bed. The Colonel would still be in his library sipping whisky, but he wouldn't hear me.

Torch in my pocket, I made two cautious journeys downstairs, piling my things outside the porch door. Retrieving my creaky old bike from the shed was far from noiseless, but no one heard and strapping my case onto the pannier, the carrier bag perched precariously on top of the basket – it wouldn't fit inside – and my shoes over the handlebars, I wobbled my way quietly, slowly, down the drive and out into the lane.

Phase One accomplished. I'd escaped!

It was raining again and even with my bike's front light on, it was hard to see. I couldn't use the torch; I needed both hands to steer the top-heavy bike. Progress was torturously slow, keeping my balance difficult. Suddenly, the bag tipped off the basket, strewing contents all over the wet road. Scrabbling to retrieve them, my bike toppled over. I fell on top, hitting my chin and howling with frustration. The headlight flickered on and off. I felt terribly alone. How *mad* was I doing this? I'd never minded cycling alone in the dark but now I was on foot aware of strange rustlings in the hedge, an owl hooting, a weird scream – was it a fox? I was scared rigid.

Managing to right myself and the bike, we wobbled on for a few yards. Then I heard a thump behind! My heart raced – but it was my case laying in the road. How had it slipped off the pannier? Exasperated, I laid the bike on its side and strapped the case on more firmly, but a few more garments fell out of the crammed-full shopping bag. Swearing like Jim, (hoping God wasn't listening), I tied my pyjama trousers, an odd sock and the woollen scarf round the dripping handlebars to join the now soggy shoes.

How far had I come? Where was I? My headlight still kept flickering on and off. I prayed it wouldn't expire, or I might. The relentless rain had seeped through my mack and overcoat already. I kept muttering, 'keep going, keep going. I *must!*'

Suddenly, I found myself on the main road into Fleet, just as deserted, just as dark and two miles still to go. Zombie-like, I staggered on, arms on fire, chilled through, my creaky old bike rattling in protest.

It must have been about three a.m. when I stumbled into the station and rolled, in a collapsed state onto a bench. A lone porter took my rail voucher, looked me over suspiciously and made some rude remark about 'bloomin' tramps' which I ignored with a stony face.

I travelled with the churns on the five a.m. milk train and at Basingstoke began my long, wobbly walk home. It was about seven o'clock. when my bike and I crumpled onto the doorstep of 27 Sarum Hill, both worn out, me starving hungry, soaked through and *so* sorry for myself.

I knocked, and knocked again, then waited for ages before the front door finally opened and there stood my utterly astonished Mother.

Never was I more pleased to see her!

* * *

TWENTY-ONE

I was now at home on 'respite' leave, organized by Mrs Bartholomew, who seemed quite agitated, sympathizing with my plight and placating the irate Fortescues at my sudden disappearance; Mrs Fortescue adding yet another sin to the list. I had left my bedroom very dusty! As if I ever had time!

With expressions that implied 'we told you so!' the chorus of concerned home voices said, 'You must leave the Land Army, it's highly unsuitable. Just look at you; tired out, much too thin, badly treated.' But my inner voice was telling me that if I left now I'd be a failure and I couldn't bear the thought. I loved the cows, the outdoor life and most of the work.

Mrs Bartholomew said she'd enquire around and see if there was another dairy farm who needed a girl. In the meantime I could enjoy home, being cosseted by Mother, having my clothes dried out, washed, ironed and mended and sipping tea in bed until eight a.m. Bliss!

It wasn't too long before Mrs Bartholomew called again to tell me that a land-girl was needed for milking on a large Estate farm, Malshanger, only about five miles away from home.

It belonged to a Sir Jeremiah Coleman and immediately I realised he must be a descendant of the famous mustard manufacturer, the nineteenth century Sir Jeremiah – and what housewife didn't have a little yellow tin of it in their kitchen cupboard.

Mrs Bartholomew had found a billet for me with one of the Estate farm workers and his wife. So despite home dissent and my nervousness, she drove me next day to meet the bailiff, Mr Kenneth Willis, and Mrs Tucker, my landlady-to-be.

After leaving Basingstoke on the Salisbury road, we eventually turned down the long, tree-lined Malshanger avenue, near Oakley, until we came to a spread of cottages and farm buildings and glimpsed the impressive big house – so different from Brinkleigh Farm.

I dreaded meeting the bailiff but the smiling man who shook my hand and greeted me warmly wasn't a bit like Jim!. He was much younger with brown wavy hair framing an amiable face and smartly dressed in cord breeches and a suede jacket.

He made us tea and while we drank it, told me my jobs would be to milk the cows, deliver it around the estate and help elsewhere when needed. My hours would be six a.m. to around six p.m. but overtime at harvesting. I would get one regular day off a week and the occasional weekend, plus Christmas this year. I couldn't believe my luck. I was relieved he didn't ask why I'd walked out of my last job but guessed Mrs Bartholomew had told him. I was to start the following week – *if* I felt ready!

I was equally nervous about meeting Mrs Tucker – would she be another Doris or Ethel? But the neat little woman who opened her cottage door was much older, her iron-grey hair marshalled into precise corrugated waves and everything about her and the room she showed us into, in neat, polished order. At her feet, a little black Scotty dog growled, until she scooped him up and said 'There, there, treasure. He don't like visitors' – which didn't bode well for me.

Introductions over, Mrs Tucker took us up to see my little bedroom, so shiny and pristine I guessed not one speck of dust would ever dare settle. I was going to miss my pretty, carpeted room and own bathroom at Brinkleigh farm. Leaping out of bed onto cold, shiny linoleum at five a.m. on a bleak winter's morning was going to take some getting used to, as were once-a-week baths again. But Mrs Tucker seemed pleasant enough so I determined to make the best of it.

While Mrs Bartholomew discussed ration books and rent with her, I tried to get friendly with 'treasure', (his real name?), but he snarled and showed his teeth. Oh dear, was this yet another animal to add to the growing list of creatures who didn't like me? Visions of Merry, Big-un and Romulus floated depressingly before my eyes. Was Treasure perhaps a reincarnation of my Granny's snappy Scotty dog who took a piece off the end of my nose when I was only two? I sincerely hoped the cows would like me.

On a mild September morning soon after, Mrs Bartholomew drove me back to Malshanger where I settled into my room. Mrs Tucker kept a tight hold onto a disapproving Treasure, explaining to him that he must be nice to me as I'd come to stay. He wasn't impressed.

At lunch time I met Mr Tucker, a friendly, well-built man who told me he mostly drove the tractors and to call him Will. The meal was plain, helpings adequate but small. By now everybody in Britain was getting used to eating even less – 'tightening our belts' was the cheerful cry – but we manual workers, despite getting extra rations were always hungry, especially me!

After lunch, Ken Willis called in his van and took me round the Estate, pointing out where the various farm buildings were, the farm workers' cottages and houses and the back entrance to the big house where I'd deliver their milk every morning.

At three-thirty he took me to a large, airy milking shed to meet the cowman, Sid Curlow and my new charges.

I stared in amazement at eight enormous cows, the colour of shiny red chestnuts, standing placidly in their stalls. How different from my dainty Jerseys girls I'd left behind!

'What are they?' I asked.

'Red Polls,' Ken told me, 'Small herd, providing milk for Sir Jeremiah and the Estate workers. Cross between a Suffolk Dun and a Norfolk Red – great milk/beef combination.'

'They's easy milkers,' Sid said. 'Good steady yields. Don't have no problems calving neither.'

'But there's no milking machine?' I suddenly noticed.

'Don't need one. Done any hand milking?' Ken asked me.

I shook my head, instantly alarmed.

'T'aint hard, miss,' Sid told me, 'soon get the hang of it, you will.'

'I'll leave you then,' said Ken. ' You can find your way back to the Tuckers.'

'Just you watch me, Miss,' said Sid, kindly, ''tis simple. I can't be doing with them new- fangled machines.'

And simple it seemed, but slow, even leisurely! Apart from udder washing and entering yields on the Min. Ag. sheets, nothing resembled the quick-action routine at Brinkleigh Farm. I watched carefully as Sid moved quietly and steadily from cow to cow, taking up to ten minutes to milk each one, then hanging the bucket on a large scale to weigh the yield. The milk then went through the dairy cooler and into a churn. The cows stood quietly, not agitating for their ration of concentrate, so I felt fairly sure they would be easy to milk with their big firm teats and mild natures. Wasn't it what I'd originally hoped for?

Although Sid told me their names, it wouldn't be easy to remember them as they *did* all look the same!

'Don' you be frettin' about names, Miss,' Sid said, smiling. 'They's each got a number, see.'

I asked if I'd have to help with the bulls.

'Oh, don' you worry, we only got one,' he said, seeing my anxious face, ' an' I sees to him, but he's a lamb most times.'

I couldn't quite believe the contrast between Jim and Sid – he reminded me of rosy-cheeked farmers I'd seen in Victorian paintings and W.L.A. posters; sturdy, benignly cheerful – and hopefully patient with novices like me! As we walked the cows back up the lane to the field they were pastured in, he chatted away amicably, asking me about machine milking and reassuring me I could 'take me time learning to hand milk.' If Jim could hear him, he'd be grinding his teeth and uttering choice, unrepeatable words!

Cleaning up after milking, which I helped Sid with, was much the same as before, but without Jim timing my every action.

The walk back to the Tuckers took me less than ten minutes, but I was finding meals there a bit fraught, as 'darling' Treasure kept escaping from the kitchen where he was imprisoned, and snapping at my ankles while we ate, giving me acute indigestion. Mrs Tucker seemed much more concerned about Treasure's feelings than my discomfort.

Four things woke me almost instantaneously next morning; a clap of thunder, my alarm, Treasure howling and bare feet on cold linoleum.

Up early with her husband, Mrs Tucker had made a pot of tea, but was sitting cuddling a shivering Treasure, who looked utterly comic wearing a pair of knitted earmuffs. I couldn't help laughing; very tactless of me!

Glaring, Mrs. Tucker said 'Awful afraid of thunder he is, poor little Treasure. Aren't you, darling? I'll have to sit here 'til the storm's over, won't I Will?'

Nodding patiently, Will strode off up the lane, while another loud clap of thunder sent me scurrying to the milking shed, just as it began to rain. The cows were already in their stalls and Sid busy setting up in the dairy. Feeling guilty, I asked if I should have arrived earlier.

He shook his head. 'Just you get the hang of milking first, Miss – I'll show 'e in a moment.'

Standing around in a milking shed without a job was so alien to my being I felt quite lost, until Sid took me over to the first cow, Gertrude. After washing her udder, he stuck a bucket underneath and settling himself on the stool, showed me the correct hand and finger positions to use, emphasizing milking front/back teats diagonally. It looked quite easy. After all, I'd stripped off Jerseys before, so I took Sid's place confidently. But oh, the disappointment! Despite my copying everything he showed me, Gertrude decided she was only going to let a dribble through occasionally, however hard I tried. Red faced, I waited for a rebuke.

'Not t' worry, Miss, early days,' Sid said, 'She knows you're new at it. Proper canny they are, these ladies. You just let me finish her, then you can have a go on Violet.'

But Violet and the other ladies were just as canny and uncooperative and by the time Sid had taken over from me for the third time, my arms ached, my hands and fingers were on fire and my confidence was in the gutter. Sid was remarkably cheerful about my failure and suggested I make a special fuss of each cow until they got used to me.

'You'll see,' he said. 'You be milkin' like an old hand come next week.'

I prayed he was right. My afternoon efforts did produce a few more teaspoons of milk from each patient lady, but by the time I'd sat through three more Treasure-disrupted meals at the Tuckers, trying not to let the escapee nip my ankles, I went to bed with a heavy lump in my stomach that was more like despondency than indigestion!

※ ※ ※

TWENTY-TWO

As Sid predicted, my milking skills did improve as day by day the docile red ladies obliged me with ever increasing quantities in my bucket, but my abilities to charm Treasure into accepting me as a housemate shrunk to zero.

Wet, windy evenings put me off exploring my new surroundings and I spent long, lonely hours in my bedroom. Tiresome Treasure had objected fiercely to my listening to the wireless downstairs, despite Mrs Tucker nursing him. He'd give a triumphant bark as I hurried out of the room.

I could tell Mrs Tucker was beginning to regret having me. I tried hard to be friendly and helpful, but suddenly she turned on me, spitefully mimicking the way I spoke. Taken aback and hurt, I vowed I'd say as little as possible, but mealtimes were miserable. Will looked embarrassed, Mrs Tucker purse-lipped and Treasure and I were now sworn enemies.

I was glad when my first day off arrived and I managed to catch a bus to Basingstoke on the main road, deciding to ride my cycle back. A mechanically-minded friend had given the poor old bone-shaker a new lease of life, so I hoped my return ride to Malshanger would be far less noisy.

At home, an official letter was waiting for me with an exciting invitation to a Women's Land Army Rally and Celebration in Southampton in the middle of September. We were to march through the City, ending up at the Guildhall square, where important V.I.P's would give speeches, followed by a special lunch and then boat trips round the harbour to view the anchored ships just returned from war zones around the world. I really wanted to go, but I'd have to summon up courage to ask Ken Willis for extra time off.

My Mother's questions about my billet I managed to tactfully deflect onto my newly acquired skill as a milkmaid, and emphasized how friendly my boss and fellow workmates were.

Returning to the Tuckers I felt awkward, but decided I'd try again to be friendly, even if I couldn't win Treasure round. Mrs Tucker said she had no room for my bike so Sid found space in a shed behind the dairy. While we were there, he pulled out a long, low wooden cart on four wheels.

'Well now, we puts the milk churn in the middle with a big jug and ladle,' he explained. 'I'll give 'e a list of doors to knock on; the ladies come out and you fills up their jugs. They knows how much, don' you worry.'

This was the week, of course, *I'd* take over from Sid and start delivering milk around the Estate, in such a unique way and *I* was the horse! Quite a relief. I couldn't wait to get started; it looked such fun.

'Ah, but first you takes up milk to the big house,' Sid told me. 'I'll show 'e how much. They makes their own butter up there.'

I was still much slower milking than Sid, but my hands ached less and Sid seemed satisfied. The red ladies and I were easy with one another now, but secretly they couldn't compare with my affectionate Jersey girls. I missed them so much.

I missed a lot of things; feeding the chickens with Masie – grinning at me through her huge glasses, fun and games with Billy and Susan, gentle Flora and little Lily and even Jim when he wasn't swearing at me. Why had everything gone so horribly wrong?

But this was a new start. Everybody here was really kind and helpful, (discounting Mrs Tucker and terrible Treasure) and I was sure I was going to be happy.

Although the wooden cart was heavier than I'd realized and it rained relentlessly, I delighted in my milkmaid duties Timidly knocking the back door of the big house, I caught glimpses of a large, decoratively tiled room with marble-topped counters.

Knocking on all the cottage doors, the wives came hurrying out with their jugs, curious to meet me, know my name, where I'd come from, talk about the weather. It took much longer than it should have that first morning, incurring Mrs. Tucker's wrath when I was late for breakfast. I hoped I'd get quicker.

Making an effort to lighten the mood, I told the Tuckers about the coming Rally in Southampton, but suddenly Mrs Tucker copied my voice again, mocking the way I spoke! Upset, not understanding why, I realised she must really dislike me. How much longer would she want me living under her roof?

Losing my billet once in a year was perhaps forgivable, but *twice*? Whatever would Mrs Bartholomew say? I dreaded to think.

***

TWENTY-THREE

Not only did Ken Willis agree to give me extra time off for the Rally, he said, 'Well, as it's on a Saturday, take it as your weekend. End of this month, come good weather or bad we must get the harvest in. It'll be all-hands-all-hours!'

He was as gloomy as Jim and the Ministry of Agriculture reports in predicting a very poor harvest – just when Britain's need was so great.

He took me with him to inspect some of the fields, pointing out wind-flattened barley and acres of wheat that should have been 'waving golden in the sunshine' but stood unripe and stunted. I was shocked, realising now how serious it was.

'Root crops are rotting in the ground,' Ken grumbled. 'Pretty sure we'll have potato rationing this winter.'

So although it seemed rather inappropriate that the Women's Land Army was having a Celebration Rally, I couldn't help feeling really excited when I stepped out of Southampton Station and found myself surrounded by other excited, high-spirited land-girls; overwhelmed too, as more and more joined us until we must have numbered hundreds. Everyone seemed to know someone else. I looked around, hoping to spot the two land-girls I'd met on my Birthday, but no luck. I felt alone and envious of the others.

Harassed-looking representatives started corralling us into orderly lines, four-a-breast, so I attached myself to the end of a row, smiling shyly at the girl next to me. In nearby streets, gaily decorated tractors and horses and carts were gathering. Somewhere up ahead a band struck up a rousing tune as we were urged forward to begin our march through Southampton streets, laughing as we tried to get in step. Three girls headed the procession, holding up a big 'Hampshire Land Army' banner. The tractors and carts filtered in between our lines. No bulls in the parade, thank goodness! The sky suddenly lightened and the sun appeared – as if in approval!

More and more onlookers were gathering, clapping and cheering. I felt quite proud, but *very* undeserving. All these hardy, cheerfully confident girls walking with me, many wearing long-service armbands, were the real heroines. I hadn't even done a year nor covered myself in glory!

Marshalled into tidy lines facing the imposing City Hall, we listened to a congratulatory message from our Patron, Queen Elizabeth; endless words of praise from the Mayor and two Land Army bigwigs. By now we were over-hot and very hungry, so the announcement to file into the banqueting hall for lunch was extremely welcome.

Hardly a banquet, but devoured with great relish, the hall reverberating to unrestrained voices of land-girls-let-loose, the friendly ones I sat with including me in their chatter.

Divided into groups of about twenty we were packed into launches to view the awesome array of destroyers, cruisers, mine sweepers and troopships moored in the harbour. One troopship had just arrived, packed with jubilant soldiers lining the rails, calling out and whistling us as we chugged by. Now we were the ones to cheer them, some girls shouting out quite shocking – unrepeatable – remarks! It was an amazing sight I never forgot, leaving me humbled and grateful to all those brave soldiers, sailors and airmen who'd won the peace for Britain.

Bursting with patriotism, I travelled back to Malshanger determined to work harder than ever, earn my long-service armband, and make yet another effort to get Mrs Tucker to like me – even if Treasure was a lost cause.

I'd saved the precious silver-paper wrapped chocolate biscuit we'd all been given, (a gift from the U.S.), and presented it to Mrs Tucker at breakfast next morning by way of a peace-offering. She looked taken aback, but muttered 'thank you' and was slightly more friendly when I told her and Will about the wonderful rally. Slowly, slowly, step by step, I told myself, keep trying.

Celebration Rally, Southampton

In the dairy that morning Sid told me Ken had started harvesting and they'd stooked up a field of barley, hoping the sun would dry it, and ripen the wheat. Alas, the sun decided to play hard-to-get and hid behind dull, rainy skies again.

By now, I was on friendly, gossiping terms with my early morning ladies-with-jugs. They gossiped, I listened, having nothing much to contribute, but my daily trundle around tugging the wooden milk-cart was one of the highlights of my day.

Occasionally I bumped into Sir Jeremiah Coleman – the first time not knowing what to do. Should I curtsy? I'd never met a Sir before. But he was such a friendly, easy-going sort of man who loved walking around his estate, taking an interest in everything that was going on. One day, I was delivering milk to the big house when Lady Coleman opened the door, very concerned if I was happy here, and at last I could say 'yes' and mean it.

Even life at the Tuckers was marginally better, and if it wasn't raining after supper I'd explore the vast estate. One evening Sid invited me to 'meet the missus and kids'. Freda came as a surprise, being quite a bit younger than Sid; thin, harassed-looking but friendly, and overwhelmed by a boisterous three- year-old boy and a large, demanding baby girl, who made my head ache by the time I left – even more than Treasure did. Freda said to call any time but I doubted I would, secretly preferring cows to small noisy children.

I was first in for breakfast at the Tuckers next morning, when suddenly the kitchen door burst open and Treasure exploded into the room. Growling, heckles up, with a threatening snarl on his mean little face, he hurled himself at me. I leapt up. Too late – he'd sunk his sharp teeth into my ankle just as I tried to get away.

Yelling 'GET OFF ME, YOU HORRIBLE DOG, GET OFF!' I shook my leg about, hoping he'd fall off. But he wouldn't let go. I slapped him on his bottom. He still clung on, growling ferociously. Kicking my leg out again, I slapped him much harder, shouting, 'GET OFF, YOU BEASTLY ANIMAL' – just as Mrs Tucker appeared, with a face that should have annihilated me on the spot.

'OOH! YOU NASTY GIRL! WHAT ARE YOU DOING TO MY LITTLE TREASURE?' she screamed. 'Come to Mummy, darling,' and she rushed over and tugged him off my ankle.

'He was *biting* me,' I protested. *'He's* the nasty one.' I rubbed my ankle. No blood showed through my thick sock. I pulled it down; luckily only deep red marks.

'You were *hitting* him – I *saw* you.' she cried. 'My poor baby – he be terrified!' Treasure whimpered obligingly in her arms.

'I told you, he went for me,' I insisted, 'and wouldn't let go.'

Will arrived at that moment, totally bewildered by our shouting match.

'She's been attacking our Treasure,' she told him, 'I'll get the police!'

'Now now, dear,' said Will, looking at my ankle, 'He do go for her a lot.'

Incensed, Mrs Tucker turned to me, 'YOU CAN GET OUT OF MY HOUSE! MISS POSH WITH YOUR LA-DE-DAH TALK AND WAYS. GET OUT – I AINT HAVING YOU HERE NO MORE!'

I just stared at her, stunned. Posh? Not me. La-de-dah? I couldn't help how I spoke.

Will was shaking his head. 'You can't turn her out, dear.'

'You keep out of it, Will.' she snapped. To me, she said, 'Pack yer things and go – THIS MINUTE!'

Up in my bedroom, I couldn't believe what had happened. Where would I go? I'd have to telephone Mrs Bartholomew. She'd be none too pleased.

Putting all my things outside the front door, I dashed across to the dairy shed, brought my bike back, and piling everything aboard manoeuvred them back to the shed – for the time being.

It was like a bad dream that kept repeating itself. If it wasn't a belligerent bull, a mean-eyed pony or a big-bully pig, it was a nasty, pint-sized dog – ruining my chances of ever becoming a successful land-girl.

****

TWENTY-FOUR

It was confession time. I couldn't put it off.

First to Sid, who said I could leave my things in the shed for the time being and he was sure Freda could find me a bite to eat for lunch.

Second, to Ken Willis, who didn't think there was anyone else on the Estate with room enough for me but he'd enquire immediately.

Third, Mrs Bartholomew. There was a weary sigh at the other end of the telephone when I told her my predicament but she said she'd drive over straight away and 'sort the situation out'. If she thought Mrs Tucker would have me back – or me return – she was quite wrong.

Fourth, Mother and the aunts, who only got to know when I arrived home that evening, unannounced and dejected. Ken had been unlucky finding me a new billet on the Estate and Mrs Bartholomew couldn't help either. I was homeless!

After the expected home chorus of 'leave the Land Army, it's highly unsuitable,' and me digging my toes in, it was agreed I should work from home. Mother was delighted, but for my aunts I could tell, a great sacrifice. Extra washing, cooking, and my lingering 'eau-de-cows' perfume they found so distasteful.

Mrs Bartholomew seemed relieved. She wouldn't have to find me any more billets. I surely couldn't get turned out of home – could I? She coped with my ration books transfer and rent payments, and I knew she hoped I wouldn't be troublesome again. The downside for me and no doubt Mother, was sharing her bed again up in our attic refuge.

So next morning, up at four forty-five a.m., and fortified by a flask of tea, and toast burned on a tiny oil stove, I cheerfully set out to cycle the five miles to Malshanger, armed with sandwiches and an apple. But the slight morning drizzle quickly turned into driving rain, nearly blinding me. A ferocious wind whipped up, almost toppling me off my bike. By the time I staggered into the milking shed, wet through, exhausted, even harbouring those tiny tempting thoughts of 'a nice little typing job just down the road', I was very late and starving hungry. What a beginning to my New Start!

The rain and wind had worked itself up into a full-blown storm so Sid decided we'd keep the red ladies indoors. How I managed to deliver the milk without the cart taking off on its own and me being spiralled up into a black hole, I'll never know – perhaps my rain-sodden clothes kept me earthed?

Sid took me home to dry off and eat my sandwiches – gobbled down fast with none left for lunch. But kind Freda insisted I shared their meal, though the sight of a baby with its face smothered in messy stew and a grubby boy peeing into the fireside log pile, rather spoiled my appetite.

The storm raged all day and Ken was deeply worried about the crops. They'd started cutting another field of barley the day before and he'd found mildew growing on the barley already stooked. Feeling pretty mildewy myself, I made my return journey home under a slightly less violent sky, to enjoy my saved, plated dinner, have a quick wash and fall into bed, sleeping so deeply my Mother's snores, for the first time ever, didn't wake me!

By the time I set out next morning the storm had died right down, leaving debris, chimneys toppled, lost tiles and fallen trees in its wake. On the farm we had some clearing up to do, but nothing major.

As if the heavens suddenly decided we'd been punished enough, especially farmers, the wind died to a whisper, the curtain of dreary clouds parted, revealing a blue sky – something we'd hungered for. September was almost over, so farmers everywhere began frantically catching up with the harvest, and at Malshanger, after milking was over and the cows put out to pasture, Sid and I joined the other workers in the fields.

They welcomed me warmly; Jolly Ned and shy, blushing Pete, ginger-haired Fred, bearded George and twinkly old Eddie and his son Dusty; mostly older men (the young workers not yet back from the war). I was immediately made to feel needed and useful.

Harvesting was in full swing. I stared in wonder at the large, clattering threshing machine exuding clouds of dust, devouring it's diet of wheat stooks being fed in by Fred, Pete and old Eddie. Ned was filling sacks with grain as it poured from the thresher's belly. At the far end, Dusty set me to work on the binder – a long, strange-looking metal contraption where the superfluous straw was compressed into large oblong shapes that came steadily chugging towards me – reminding me of a giant sausage-making machine! Given a pile of wire lengths, my job was to securely tie each bale round twice as they were disgorged. Will, smiling sheepishly at me, with George, loaded them onto carts ready to be pulled away by two magnificent shire horses standing patiently nearby.

I found twisting the wires quite painful wrist and finger work – but so what! The sun was gloriously warm, the men's banter friendly and cheerful and even the rejoicing birds made themselves heard above the noisy thresher. I relished the feeling of us all working so happily together, imagining the thresher as a fat Queen bee diligently attended to by her numerous, ever-busy workers.

As a change from twisting wires, I was asked to lead one of the huge horses, cart piled high with straw bales, down the lane to a big barn for storage. I was scared – memories of Merry still haunted me – and this great beast towered above me. Would its giant hooves crush my feet? Would it obey me when we had to turn left towards the barn? My hand trembling on the bridle, I quaked 'walk on' and prayed hard.

I might just as well have had a label on me 'NOT WANTED ON VOYAGE'! Not only did my feet stay intact, my gentle giant took complete control, leading me down the lane, turning left just at the right moment, stopping at the exact spot to be unloaded, then leading me back up the lane to where we'd started. It was the easiest job I'd ever done. I made the most of my leisurely strolls, but alas, I still had to do my share of painful wire twisting on the baler.

VICTORY HARVEST

You are needed in the fields!
APPLY TO NEAREST EMPLOYMENT EXCHANGE FOR LEAFLET & ENROLMENT FORM OR WRITE DIRECT TO THE DEPARTMENT OF AGRICULTURE FOR SCOTLAND 15 GROSVENOR STREET, EDINBURGH.

We worked until dusk every evening, moving from field to field, with short breaks to gulp down water, cold tea and buns, then carry on. I'd cycle home white with dust, every inch of my skin itching with chaff, to stand outside the back door and peel off and shake almost every garment before I could step inside. Only then could I pacify my growling stomach and feel human again.

A day off at last, Sunday, and luckily the Harvest Festival service at St.Michael's church. Smartly dressed in my uniform,

I joined the packed congregation, admiring the flowers and piles of vegetables and fruit stacked around the pulpit – *plus* two stooks of wheat!

As we lustily sang 'We plough the fields and scatter...' I swelled with pride, sincerely hoping the congregation was appreciating how much I – *we* land-girls – had done to gather in the harvest.

I chose to ignore the Angelic voice whispering in my ear 'thou shalt not boast nor have false pride' and basked in my own glory. Just for the day.

****

TWENTY-FIVE

With some overtime swelling my pay packet, I felt very rich but didn't have enough coupons for a new jumper, so I decided to save up for Christmas.

October days were still warm and dry, so Sid and I did some Autumn cleaning-out jobs, and treated the red ladies to extra grooming, now a pleasure for me to milk and look after, even though I missed my Jersey girls and pined for Josephine.

Soon after, Ken asked me if I'd like to help him prepare some root crops for winter feed for the large flock of sheep kept at Malshanger. *Asked* me – not told! I still couldn't believe how kind and thoughtful everyone was, not counting Mrs Tucker and tiresome Treasure

In one of the barns, Ken showed me a huge heap of turnips and mangels piled in a corner, and in the middle stood what looked like a giant mincing machine – a huge replica of the one my aunt screwed to her kitchen table to mince up the Sunday joint remains for Monday's cottage pie!

Powered by a petrol engine, the mincer was certainly noisy as Ken and I shovelled big spades-full of the heavy turnips and mangels into its yawning mouth, to immerge as small chopped-up lumps. Ken said this would be an ongoing job throughout the winter and although my back and neck complained, I enjoyed Ken's company. He was the first person who'd ever talked to me like a grown-up. Telling me about farming, asking my opinion, discussing politics – of which I knew lamentably little – and making me feel what I said was important.

With the evenings drawing in, I was home earlier and revelled in going to the cinema with my friends on my days off. The aunts still flung the windows wide open when I was around and decided

I needed extra baths. Mother got a part-time job at a baker's shop and if by closing time there were any buns left over, staff could buy them. Oh, the delight in devouring those bright yellow buns made from dried eggs, no fat, the odd current or two and hard as rocks – but when you're hungry – ! There was no easing up of rationing. Mr Attlee kept telling us most of Europe was still very hungry and we had to help the Americans feed them. Mrs Bartholomew said the Women's Land Army wouldn't be disbanded until we weren't needed; some time yet, she thought.

A large photo of the Rally arrived in the post. There I was, nearest row, five back; looking like I belonged, feeling I had a purpose, though still wistful I wasn't working with other land-girls; it would have been fun.

But how orderly and peaceful my life had become. No eagle-eyed Jim swearing at me, timing my every move. No dramas with unfriendly animals and never expected to do the impossible – and yet!

It was Ken's interesting chats about farming that first made me think about making farming my career. The W.L.A. may be disbanding but I could carry on. Why not?

He suggested I join the Young Farmers Club in Basingstoke; get to know like-minded others. I couldn't attend all the meetings but found I knew a few ex-school mates, sons and daughters of local farmers. Being the only land-girl – and one of the few actual hands-on members – I felt a certain secret superiority and took down copious notes at lectures. There was certainly a tremendous lot to learn about farming.

I was invited to spend one of my day's off on a Farmer's Club member's dairy farm and help milk their Shorthorn cows. It was a different but interesting experience; the milking machine simpler than at Brinkleigh farm, the cows large and stocky like Alice. I definitely preferred their dainty cousins.

I saw myself in the distant future, owning my own dairy farm. Jerseys, of course – one named Josephine. I guess I'd have to have a bull, but someone else could deal with him. There'd be lots of chickens, *no* pigs and I'd grow wheat and barley which I'd broadcast by hand, and make silage. Maybe eventually I'd marry a farmer – or not. I had still given up on romance.

In the meantime all I had to do was work hard, keep learning and have my sights firmly fixed on my goal.

What a good idea it was, I decided, to plan out one's life in advance. You just had to be patient and your dreams were bound to come true!

***

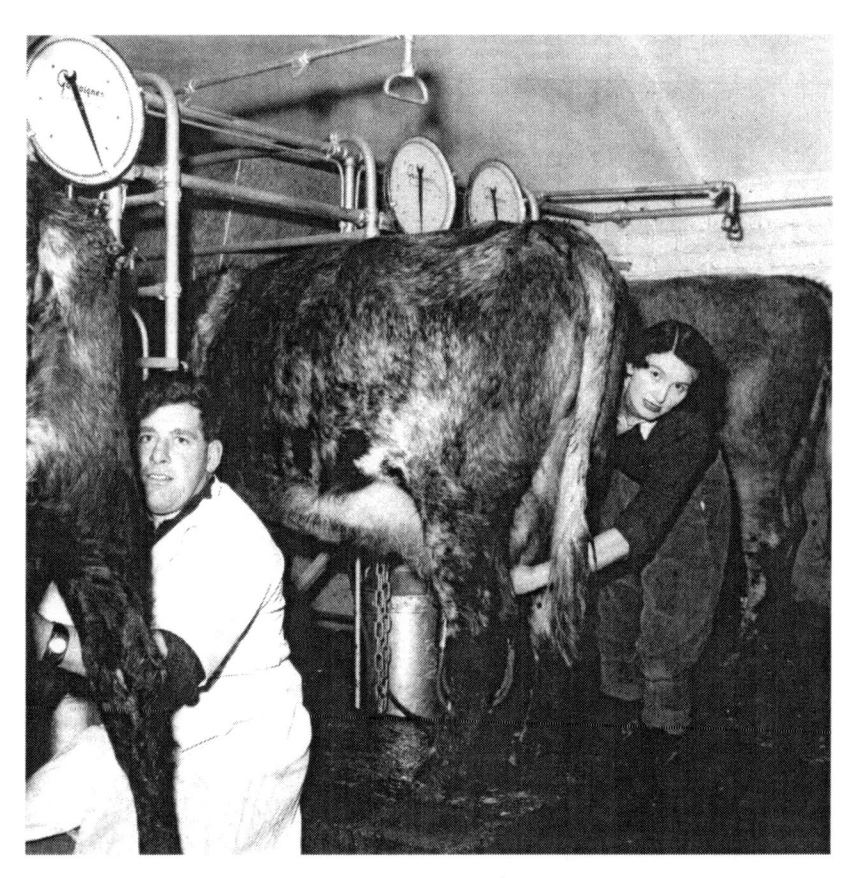

helping milking on a neighbouring farm

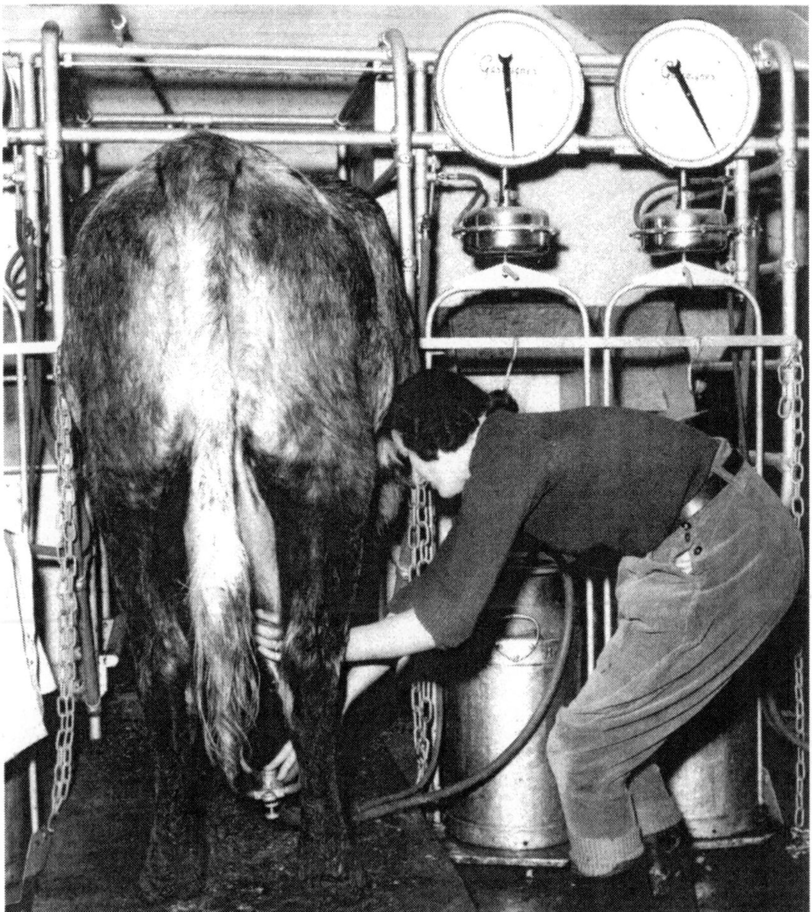

TWENTY-SIX

My state of euphoria lasted just five days. Reality began seeping in. Everything had been going along so happily; October days dry and mild; the harvest, such as it was, now safely gathered in and my notebook rapidly filling with all I'd need to know to become a successful farmer.

With my mind full of positive thoughts, I was cycling to work early one morning when the ominous clanking noise I'd detected a few days earlier got significantly louder. Suddenly, the bike juddered, I heard the chain noisily detach itself and we veered into the hedge. A peddle fell off! I looked frantically around for help but the road was deserted.

Two more miles to Malshanger! I had no option but to push the bike, trailing its errant chain, the peddle in my basket, and hope a car or lorry would pass me, but nothing did. I arrived very late and shaky just as Sid was finishing milking..

He fixed my chain and peddle back on, but shaking his head said, 'Seen it's last days, I reckon.'

My spirits descended to my gumboots. No bike – no job. I couldn't afford to replace it. My £4.11s.3d savings wouldn't go very far. A friend had just paid £16 for a new bike! Mother couldn't afford to help me either.

My mechanically-minded friend reluctantly gave the bike yet another once-over, tightening nuts and oiling it, then repeating the obvious, 'Why not buy yourself a new one?' As if! My bike had to last. Buses were infrequent, none very early or late evening. I had no option but to cycle on hopefully.

There was extra work now, the cows needing daily hay and silage feeds as winter approached. With milking, the daily trundle around the estate and helping Ken mince root crops, each day was action-packed and flew by. It seemed to be work, cycle home, dinner, wash, bed, sleep like a dead thing, up early, cycle to work, more work …. ad infinitum! I had less time for a social life now, even though I lived at home.

October's mild autumnal days were slowly overtaken by November chills with rain and occasional gales. Creeping down the dark attic stairs in the early hours each morning and battling all weathers on my bike-on-borrowed-time was much more of an effort, and home voices murmuring 'You'd be much better off in a nice office job, dear,' didn't help. But I had my sights firmly fixed on my golden future and providing my bike held up, nothing could deter me.

Christmas loomed and the promised two days off arrived just in time to decorate a small tree, wrap up presents and help make mince pies. The Food Ministry had allocated each household a little extra fat, sugar and dried fruit to cook Christmas fare. A big parcel from my sister in America was full of wonderful treasures: tins of spam and butter, dried fruit, chocolate and cheese, plus gifts. For me, a black velvet skirt, the pocket decorated with tiny pearls. I was ecstatic, though the aunts said it was much too sophisticated for my age.

Spoiled by Mother, allowed my thimble of sherry to toast our King, George V1, carols in church, the annual family card game played for halfpennies, and feeling like a princess in my beautiful velvet skirt at the Boxing Day party, made Christmas 1946 extra special.

The Ministry of Food gave us a late Christmas present – potato rationing – so we tightened our belts even more!

I knew just what New Year resolutions I'd be making on the 31st December and was sure 1947 would see the beginning of my dream coming true.

Freezing fog not only dampened my enthusiasm for returning to the daily grind but also my dreams. The time from now to my ultimate goal seemed to stretch further and further into the distance, with such a lot of slog and discomfort in between.

But what weakness of character! Ashamed, I vowed I would keep on going, whatever – and hopefully my precious bike would too.

January added driving sleet to the daily misery and when Ken said he'd like me to join Pete and old Eddie in hoeing some turnips, alarm bells rang! Putting turnips through the mincer was quite bearable – but *hoeing* them! I hadn't forgotten the painful consequences last time. A vision of Dr. Blunt leaning over me brandishing the huge 'horse-sized' syringe and saying 'this is going to hurt,' flashed into my mind.

But I was tougher now; not the 'lily-livered land-girl' Jim often called me. So of course, I cheerfully agreed and picking up a hoe, presented myself to old Eddie and Pete in a field of unending rows of turnips.

'We be glad to see thee, Jeannie girl,' said Eddie, 'aint we, Pete?'

Pete blushed to the very tips of his large ears and grunting, quickly began hoeing in the nearest row.

'Tek no notice,' Eddie whispered. 'He be shy, I reckon. Now, you get stuck in 'bout six rows down from me.'

'Stuck' was just the word to describe my efforts to pull the hoe out once thrust into the heavy, frosty soil, but I persevered and got into some sort of rhythm, determined to ignore any niggling back and shoulder aches.

After all, it was a question of mind over matter – or rather body. Our P.T. school teacher used to say briskly 'Get out onto the hockey field and you'll soon feel better,' even if we told her we were dying.

Glancing round, I was dismayed to see Pete was just beginning a new row and old Eddie, despite being half my height and four times as old, was just finishing his. I wasn't even half way down mine.

'Never you mind, ducks.' shouted Eddie, 'You aint used to it like we be, is she, Pete?'

I could see Pete blush again even at this distance. With a wave, I plied my hoe with new vigour, determined to get faster, despite the hoe feeling heavier and heavier, as sleet turned into stair-rod rain and began soaking through my greatcoat.

When Eddie shouted 'Grub time, girlie!' I could hardly uncurl, but I'd managed to quicken my work-pace and proudly surveyed the three-and-a-half well-hoed rows I'd achieved.

While Eddie and Pete went home to their dinners, I crawled up to the dairy to dry off, gobble my sandwich and fall asleep on some sacking, until Sid woke me and it was time to put in a couple more hours' hoeing before afternoon milking.

Sir Jeremiah gave me a hearty wave as he leaned on the gate watching us and I vowed when I had my own farm, I'd delegate these disagreeable jobs, like he could. But this was now, so I worked on determinedly, longing for milking time.

'You's a gritty little heifer, Jeannie,' said Eddie, rewarding me with a hairy-looking sweet from his pocket. 'T'aint easy for you town girls t'take t'country ways, is it, Pete?'

Pete, scarlet as ever, gave me a half-grin. Progress!

Next morning, bolstered by praise but still chanting my mantra 'mind over body', I continued to do battle with the unyielding earth. I was distracted by a long, fat worm wriggling on my hoe.

'You're the culprit,' I told it, feeling resentful. 'If you did your work properly aerating the soil, it wouldn't be half so heavy.'

I only just subdued a temptation to slice it in half.

A strong wind blew rain into my face. My wool gloves were saturated so I peeled them off frozen fingers. The soaking wet greatcoat chilled my back and my scarlet nose wouldn't stop running. I just hoped the wretched turnips appreciated the loving care I was giving them in this foul weather.

But now I couldn't pretend a grim battle wasn't going on between mind and body. From my toes to my finger-tips, every joint was on fire, my neck and chest had almost seized up and my back was complaining bitterly. Impossible to ignore, try as I might.

I must rest, I told myself. Glancing round at Eddie, who worked without ever stopping, I could see why. He was much nearer the ground than me so didn't have so far to bend. If only I was shorter!

With another hour still to go, I plunged my hoe into the ground again, but just as I did, a sudden, searing spasm of pain clamped itself round my hips. I crumpled to the ground.

'OH NO!' I wailed. '*NO!*' I tried to get up, but ferocious pain kept my chin locked to my knees.

I heard Eddie calling 'What's up, girlie?' and then he was leaning over me, while I moaned and fought back tears of pain and frustration.

'You's hurt yerself?' he asked. 'Let's help thee up. Give us yer hand.'

'I can't,' I cried. 'My back's gone.'

Why did this have to happen again? I felt really angry.

'Oh dearie me, Jeannie – what'll us do?'

He called Pete over and together they managed to half carry, half drag me, still bent in two, across the field to prop me up by the gate.

'Poor little heifer,' sympathized Eddie, as I crouched in the grass. 'You rest here. Pete, go for the gaffer – get a move on!'

It seemed ages before Pete came back with Ken, who took one concerned look and went back to get his car. He drove me back to his office, still doubled up, and phoned Mrs Bartholomew. I imagined her big sigh, but she said she'd drive over as soon as possible.

I was in the depths of misery. I'd let Sid down with the milking. I'd let Ken down, and even Mrs Bartholomew, but worst of all, I was letting myself down. What about my golden future if this kept happening?

It was all a painful blur, being levered into Mrs. Bartholomew's car, driven home to pick up my shocked Mother and taken on to see the doctor.

I was hoisted back into Mrs Bartholomew's car an hour or so later, with my left wrist strapped up, two fingers in splints, a box of Aspirin, a bottle of Wintergreen rub to ease my back pain, two sticks and a letter of referral to an Orthopaedic specialist.

A week later I shuffled into the hospital on my sticks and crouched before the great man himself, waiting deferentially for him to look up from the letter he was reading about me.

He slowly shook his head.

'Who in God's name let *you* loose on the land?' he said, looking me up and down. 'You're all the wrong shape and size! Much too tall, much too skinny. It's the short, sturdy females who are made for the job.'

It was ironic – my ramshackle old bike had managed to survive, but I hadn't. Defeated by a field of turnips!

The Women's Land Army granted me an 'honourable' discharge and our Patron, Queen Elizabeth took time off from her other royal duties to send me a signed Certificate thanking me for my 'loyal, devoted service and unsparing efforts' – all very true but woefully short-lived.

I'm sure Mrs Bartholomew gave one enormous sigh of relief at my departure, as did my Mother and three aunts.

Two months later, walking very slowly on one stick and smelling of Wintergreen, with my still painful left wrist strapped and two knuckles bandaged, I started that nice little shorthand-typing job at the Engineering firm just down the road!

****

By this personal message I wish to express to you

Miss J.M. Parsons

my appreciation of your loyal and devoted service as a member of the Women's Land Army from 5. 11. 45. to 4. 1. 47 Your unsparing efforts at a time when the victory of our cause depended on the utmost use of the resources of our land have earned for you the country's gratitude.

Elizabeth R

AFTER THOUGHTS

So, my dream of earning a W.L.A. long-service armband and eventually becoming a farmer owning a herd of Jersey cows – one called Josephine – had been cruelly snatched away.

Despite the war being over for nearly two years, Britain and most of Europe were still hungry, so women continued to be needed on the land. It wasn't until 1950 that the Women's Land Army was disbanded, but with no official recognition for the essential part they had played in keeping Britain fed. Rationing didn't finally end until 1954.

It was to be 2008 before the Government did at last acknowledge the W.L.A's vital contribution, with formal thanksgiving services and a badge of honour for each of the *surviving* land-girls!

In the meantime, I didn't manage to elevate myself very far above the lowly typing pool I started out in; most certainly not reaching the dizzying heights of becoming the Prime Minister's personal secretary.

But I never forgot my first, most cherished dream, and many years later an unexpected opportunity opened up. I went to college to study art, and discovered the multiple joys of all things artistic and creative.

Which proves it's a very good idea to have a few spare dreams tucked away for a rainy day – you never know when you might need one!

****

Women's Land Army   *Women's Timber Corps*

I am delighted that so many former members of the Women's Land Army and the Women's Timber Corps are here today in order that we may thank you for all that you did for the country during the Second World War and beyond. The Women's Land Army was founded during the First World War amid fears of food shortages as male labourers joined the Armed Forces. More than 260,000 women enlisted. During the Second World War 80,000 women left their homes and families to live in lodgings and hostels across the countryside to work on farms. Wearing a uniform of green jerseys, brown breeches and brown felt hats, they kept the country fed, milking cows and mucking-out pigs, driving tractors and steering ploughs, sowing seeds and harvesting crops. Hampshire was one of the largest employers in the country. In 1942 a sister organisation was formed called the Women's Timber Corps or 'Lumber Jills' and a further 6,000 women were sent to work in forests. The Women's Land Army and Women's Timber Corps continued to serve until 1950.

In July 2008 the Government announced that it would award surviving former members of the Women's Land Army and the Women's Timber Corps a specially designed badge of honour in recognition and commemoration of their contribution during the Second World War and its aftermath. Over 30,000 applications have been received and over 1,000 of these were from Hampshire.

Mary Fagan,

Mrs Mary Fagan JP
Her Majesty's Lord-Lieutenant of Hampshire

2008 – Thanks at last!

Rate — civil liberties
scope.